London Mathematical Society Lecture Note Series. 104

Elliptic Structures on 3-Manifolds

C.B. THOMAS
Department of Pure Mathematics and Mathematical Statistics
University of Cambridge

CAMBRIDGE UNIVERSITY PRESS
Cambridge
London New York New Rochelle
Melbourne Sydney

CAMBRIDGE UNIVERSITY PRESS
Cambridge, New York, Melbourne, Madrid, Cape Town, Singapore, São Paulo

Cambridge University Press
The Edinburgh Building, Cambridge CB2 8RU, UK

Published in the United States of America by Cambridge University Press, New York

www.cambridge.org
Information on this title: www.cambridge.org/9780521315760

First published 1986
Re-issued in this digitally printed version 2008

A catalogue record for this publication is available from the British Library

Library of Congress Cataloguing in Publication data

Thomas, C.B. (Charles Benedict)
Elliptic Structures on 3-Manifolds.
(London Mathematical Society lecture note series; 104)
Notes from lectures given at the University of Chicago in Apr & May 1983
1. Three manifolds (Topology) I. Title. II. Title: Elliptic
structures on three-manifolds. III. Series
QA613.2.T47 1986 514'.223 85-12742

ISBN 978-0-521-31576-0 paperback

LONDON MATHEMATICAL SOCIETY LECTURE NOTE SERIES

Managing Editor: Professor J.W.S. Cassels, Department of Pure Mathematics and Mathematical Statistics, 16 Mill Lane, Cambridge CB2 1SB, England

The books in the series listed below are available from booksellers, or, in case of difficulty, from Cambridge University Press.

4 Algebraic topology, J.F.ADAMS
5 Commutative algebra, J.T.KNIGHT
8 Integration and harmonic analysis on compact groups, R.E.EDWARDS
11 New developments in topology, G.SEGAL (ed)
12 Symposium on complex analysis, J.CLUNIE & W.K.HAYMAN (eds)
13 Combinatorics, T.P.McDONOUGH & V.C.MAVRON (eds)
16 Topics in finite groups, T.M.GAGEN
17 Differential germs and catastrophes, Th.BROCKER & L.LANDER
18 A geometric approach to homology theory, S.BUONCRISTIANO, C.P.ROURKE & B.J.SANDERSON
20 Sheaf theory, B.R.TENNISON
21 Automatic continuity of linear operators, A.M.SINCLAIR
23 Parallelisms of complete designs, P.J.CAMERON
24 The topology of Stiefel manifolds, I.M.JAMES
25 Lie groups and compact groups, J.F.PRICE
26 Transformation groups, C.KOSNIOWSKI (ed)
27 Skew field constructions, P.M.COHN
29 Pontryagin duality and the structure of LCA groups, S.A.MORRIS
30 Interaction models, N.L.BIGGS
31 Continuous crossed products and type III von Neumann algebras,A.VAN DAELE
32 Uniform algebras and Jensen measures, T.W.GAMELIN
34 Representation theory of Lie groups, M.F. ATIYAH *et al.*
35 Trace ideals and their applications, B.SIMON
36 Homological group theory, C.T.C.WALL (ed)
37 Partially ordered rings and semi-algebraic geometry, G.W.BRUMFIEL
38 Surveys in combinatorics, B.BOLLOBAS (ed)
39 Affine sets and affine groups, D.G.NORTHCOTT
40 Introduction to Hp spaces, P.J.KOOSIS
41 Theory and applications of Hopf bifurcation, B.D.HASSARD, N.D.KAZARINOFF & Y-H.WAN
42 Topics in the theory of group presentations, D.L.JOHNSON
43 Graphs, codes and designs, P.J.CAMERON & J.H.VAN LINT
44 Z/2-homotopy theory, M.C.CRABB
45 Recursion theory: its generalisations and applications, F.R.DRAKE & S.S.WAINER (eds)
46 p-adic analysis: a short course on recent work, N.KOBLITZ
47 Coding the Universe, A.BELLER, R.JENSEN & P.WELCH
48 Low-dimensional topology, R.BROWN & T.L.THICKSTUN (eds)
49 Finite geometries and designs,P.CAMERON, J.W.P.HIRSCHFELD & D.R.HUGHES (eds)
50 Commutator calculus and groups of homotopy classes, H.J.BAUES
51 Synthetic differential geometry, A.KOCK
52 Combinatorics, H.N.V.TEMPERLEY (ed)
54 Markov process and related problems of analysis, E.B.DYNKIN
55 Ordered permutation groups, A.M.W.GLASS
56 Journêes arithmêtiques, J.V.ARMITAGE (ed)
57 Techniques of geometric topology, R.A.FENN
58 Singularities of smooth functions and maps, J.A.MARTINET
59 Applicable differential geometry, M.CRAMPIN & F.A.E.PIRANI
60 Integrable systems, S.P.NOVIKOV *et al*
61 The core model, A.DODD
62 Economics for mathematicians, J.W.S.CASSELS
63 Continuous semigroups in Banach algebras, A.M.SINCLAIR

64 Basic concepts of enriched category theory, G.M.KELLY
65 Several complex variables and complex manifolds I, M.J.FIELD
66 Several complex variables and complex manifolds II, M.J.FIELD
67 Classification problems in ergodic theory, W.PARRY & S.TUNCEL
68 Complex algebraic surfaces, A.BEAUVILLE
69 Representation theory, I.M.GELFAND et al.
70 Stochastic differential equations on manifolds, K.D.ELWORTHY
71 Groups - St Andrews 1981, C.M.CAMPBELL & E.F.ROBERTSON (eds)
72 Commutative algebra: Durham 1981, R.Y.SHARP (ed)
73 Riemann surfaces: a view towards several complex variables,A.T.HUCKLEBERRY
74 Symmetric designs: an algebraic approach, E.S.LANDER
75 New geometric splittings of classical knots, L.SIEBENMANN & F.BONAHON
76 Linear differential operators, H.O.CORDES
77 Isolated singular points on complete intersections, E.J.N.LOOIJENGA
78 A primer on Riemann surfaces, A.F.BEARDON
79 Probability, statistics and analysis, J.F.C.KINGMAN & G.E.H.REUTER (eds)
80 Introduction to the representation theory of compact and locally compact groups, A.ROBERT
81 Skew fields, P.K.DRAXL
82 Surveys in combinatorics, E.K.LLOYD (ed)
83 Homogeneous structures on Riemannian manifolds, F.TRICERRI & L.VANHECKE
84 Finite group algebras and their modules, P.LANDROCK
85 Solitons, P.G.DRAZIN
86 Topological topics, I.M.JAMES (ed)
87 Surveys in set theory, A.R.D.MATHIAS (ed)
88 FPF ring theory, C.FAITH & S.PAGE
89 An F-space sampler, N.J.KALTON, N.T.PECK & J.W.ROBERTS
90 Polytopes and symmetry, S.A.ROBERTSON
91 Classgroups of group rings, M.J.TAYLOR
92 Representation of rings over skew fields, A.H.SCHOFIELD
93 Aspects of topology, I.M.JAMES & E.H.KRONHEIMER (eds)
94 Representations of general linear groups, G.D.JAMES
95 Low-dimensional topology 1982, R.A.FENN (ed)
96 Diophantine equations over function fields, R.C.MASON
97 Varieties of constructive mathematics, D.S.BRIDGES & F.RICHMAN
98 Localization in Noetherian rings, A.V.JATEGAONKAR
99 Methods of differential geometry in algebraic topology, M.KAROUBI & C.LERUSTE
100 Stopping time techniques for analysts and probabilists, L.EGGHE
101 Groups and geometry, ROGER C.LYNDON
102 Topology of the automorphism group of a free group, S.M.GERSTEN
103 Surveys in combinatorics 1985, I.ANDERSEN (ed)
104 Elliptical structures on 3-manifolds, C.B.THOMAS
105 A local spectral theory for closed operators, I.ERDELYI & WANG SHENGWANG
106 Syzygies, E.G.EVANS & P.GRIFFITH
107 Compactification of Siegel moduli schemes, C-L.CHAI
108 Some topics in graph theory, H.P.YAP
109 Diophantine analysis, J.LOXTON & A.VAN DER POORTEN (eds)
110 An introduction to surreal numbers, H.GONSHOR

CONTENTS

INTRODUCTION

It has long been conjectured that if the finite group G acts freely on the standard sphere S^3, then the action is topologically conjugate to a free linear action. Equivalently the orbit space S^3/G is homeomorphic to a manifold of constant positive curvature, and such elliptic 3-manifolds are classified in terms of the fixed point free representations of G in SO(4), see the book by J. Wolf [Wo] for example. The purpose of these notes is to collect together the evidence in favour of this conjecture at least for the class of groups G which are known to act freely and linearly in dimension three. The main result is that if G is solvable and acts freely on S^3 in such a way that the action restricted to all cyclic subgroups of odd order is conjugate to a linear action, then the action of G is conjugate to a linear action. This reduction to cyclic groups (which is false in higher dimensions) depends on (a) the algebraic classification of the fundamental groups of elliptic manifolds and (b) geometric arguments due to R. Myers and J. Rubinstein classifying free Z/2 and Z/3-actions on certain Seifert fibre spaces. The proof is contained in Chapters I-IV; for part (a) we follow an unpublished joint manuscript with C.T.C. Wall [Th-W], and for part (b) the original papers [My] and [R2]. Besides the reduction theorem already quoted the argument implies that the original conjecture holds for groups G whose order is divisible by the primes 2 and 3 only. For the non-solvable group $SL(2,F_5)$ the corresponding reduction theorem is weaker - a free action which is linear on each element embeds in a free linear action on S^7 (see Chapter VI).

The other topics which we consider are the classifying map $B\phi: BG \to BDiff^+S^3$ associated with a free smooth action by G, the homotopy classes of finite 3-dimensional Poincaré complexes with finite fundamental group, and (in an appendix) Heegard decompositions of genus 2 for elliptic manifolds. In a "concluding unscientific postscript" we suggest various ways in which the remaining core problem of free actions by cyclic groups may be approached - but the actual results we obtain are very weak.

These notes are based on a course of lectures which I gave at the University of Chicago in the spring of 1983, and have been available in a preliminary version for some time. Among those who listened to me then I am particularly grateful to Peter May and Dick Swan for their helpful comments. I would also like to thank Terry Wall for teaching me over the years much of the mathematics on which this work is based, and for being always willing to listen to my ideas however haltingly expressed. Finally I would like to thank the Editor of the LMS Lecture Notes for agreeing to accept an expanded version of the Chicago notes for publication in the series, David Tranah of Cambridge University Press for his advice and patience, and Gwen Jones for typing the manuscript.

Cambridge, May 1986.

CHAPTER I: SEIFERT MANIFOLDS.

Let M^3 be a compact, connected 3-dimensional manifold without boundary. Where necessary we shall assume that M^3 has a smooth structure - there is no loss of generality in doing so, since M^3 is triangulable and the obstructions to smoothing vanish. Consider first a smooth action by the compact group $SO(2) = S^1$ on M^3. We use the notation

$$G \times M \to M, \quad x \longmapsto gx,$$

subject to the conditions (i) $g_1(g_2x) = (g_1g_2)x$, (ii) $1x = x$ and (iii) if $gx = x$ for all $x \in M$, then $g = 1$. Under condition (iii) the action of G is said to be effective. The orbit $Gx = \{gx: g \in G\}$ is homeomorphic to the homogeneous space G/G_x, where $G_x = \{g \in G: gx = x\}$ is the isotropy group of x. Since G is abelian, G_x is the isotropy group of each point of the orbit, and $\bigcap_{x \in M} G_x = \{1\}$ by condition (iii). The space of orbits $M^* = M/_G$ is a 2-dimensional manifold with respect to the quotient topology; the discussion of isotropy below will make this plain. Since G acts on the tangent space to x via the differential there is a representation of G_x on the normal space to x, which may be identified with the complement in T_x to the tangent space along the orbit. With respect to some equivariant Riemanian metric let V_x be the unit disc of this "slice" representation space. The equivariant classification theorem below for pairs (M^3, S^1) depends on two classical results from equivariant topology, see for example [J]:

THEOREM 1.1 <u>The total space of the disc bundle</u> $G \times_{G_x} V_x$ <u>is equivariantly diffeomorphic to a G-invariant tubular neighbourhood of the orbit</u> Gx <u>in</u> M, <u>under the map</u> $[g,v] \mapsto gv$, <u>and the zero section</u> $G/_{G_x}$ <u>maps to the orbit</u> Gx .

THEOREM 1.2 (<u>stated for abelian transformation groups</u>). <u>Let</u> G <u>act smoothly and effectively on the connected manifold</u> M. <u>Then there is a subgroup</u> $H \hookrightarrow G$ <u>such that the union of the orbits with</u> H <u>as isotropy subgroup forms a dense subset of</u> M. <u>Furthermore the orbit space of these so called principal orbits is connected</u>.

H is called the principal isotropy subgroup; the union $M_{(H)}$ of the principal orbits has the structure of a fibre bundle. The first theorem depends on the choice of an equivariant Riemannian metric on M, which gives rise to an exponential map of maximal rank near the zero section $G/_{G_x}$ of the normal bundle. Since the manifold M is compact, it is enough to prove Theorem 1.2 for the submanifold $G \times_{G_x} V_x$. Here the principal orbits belong to the complement of the zero section, a point is moved in the direction of Gx by G, and in the normal direction by G_x (modulo the kernel of the slice representation).

For the pair (M^3,S^1) we see that a closed subgroup is either $\{1\}$, S^1 or isomorphic to the finite cyclic group $Z/_\mu$. The principal orbit type equals $\{1\}$, M^* is a 2-manifold, in general with boundary. However we shall restrict attention to the case when $M^* = \emptyset$, when M^* is characterised by the pairs (o_1,g) or (n_1,g). The first symbol distinguishes between orientable and non-orientable; the second is the genus. The assumption that $M^* = \emptyset$ eliminates discussion of (a) fixed points (isotropy subgroup equals S^1) and (b) $G_x = Z/_2$ with the slice action equal to reflection about an arc. Theorem 1.1 shows that the exceptional orbits map to a finite union of r distinct points in M^*.

Consider an exceptional orbit, $G_x = Z/_\mu$ with $\mu > 1$ and case (b) excluded. Identify a slice with the 2-disc D^2, and let $\zeta = 2\pi/_\mu$ act via

$$\zeta(r,\theta) = (r,\theta+\nu\zeta), \text{ where } (\nu,\mu) = 1 \ \& \ 0 < \nu < \mu.$$

The action inside a small tubular neighbourhood N of the orbit can now (following 1.1) be written as $(r,\theta,\psi) \quad (r,\theta + \nu\zeta,\phi + \mu\psi)$, where ψ denotes the coordinate on S^1. The exceptional orbit itself corresponds to $r = \theta = 0$ and has isotropy group of order μ. The action on N is completely determined by the

<u>Seifert invariants</u> (α,β), where $\alpha = \mu$, $\beta\nu \equiv 1 \pmod{\alpha}$ and $0 < \beta < \alpha$. Changing the orientation of the pair $(N,\partial N)$ while keeping the S^1-action fixed replaces the pair (α,β) by $(\alpha,\alpha-\beta)$. With computation of the fundamental group in mind pick out a curve q in ∂N, which is orthogonal to a principal orbit h on ∂N, and given by

$$q = \{(r,\theta,\phi): r = 1.\ \theta = \rho\chi,\ \phi = \beta\chi.\ 0 \le \chi < 2\pi\}.$$

Here orientation is according to decreasing χ, $\rho = (\beta\nu-1)/\alpha$, and we note that the curve $m = \alpha q + \beta h$ is null-homotopic in the solid torus N, see diagram.

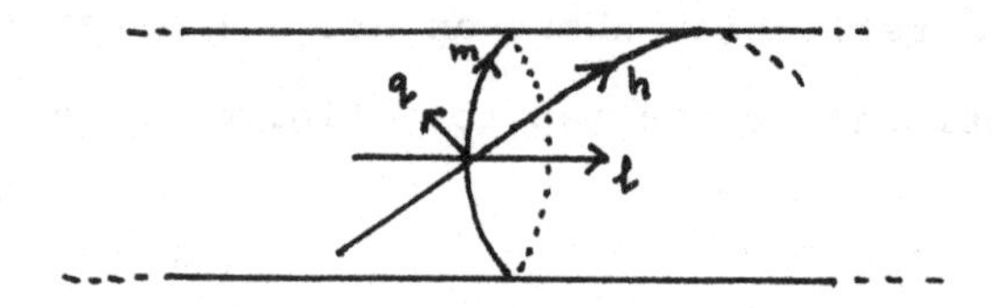

Remark: if no particular orientation of N is specified, there is an ambiguity in the Seifert invariants, unless we take $0 < \beta \le \frac{\alpha}{2}$.

When there are no exceptional orbits, that is the S^1-action is principal, the bundle $M \to M^*$ is classified by an "obstruction or first Chern class" $b \in H^2(M^*,Z)$ Z or $Z/_2$, depending on whether M^* is orientable or non-orientable. Again with the computation of π_1 in mind one can interpret b

geometrically as follows:- write $M_o = M - \mathring{N}_o$ and let q_o be a cross-section (as above) to the action on the boundary. Then with a suitable orientation convention the equivariant sewing of M_o to N_o along ∂N_o is determined up to equivariant diffeomorphism by making $q_o + bh$ into a meridian curve, that is one null-homotopic in N_o. In the general case one defines $b \in H^2(M^*-(D_1^2 \cup \dots \cup D_r^2)^\circ, S_1^1 \cup \dots \cup S_r^1, Z)$.

THEOREM 1.3 *Let S^1 act effectively and smoothly on a closed, connected, compact C^∞-manifold M^3, and assume that the orbit manifold M^* is without boundary. Then up to equivarient diffeomorphism M^3 is determined by the orbit invariants*

$$\{b;\ (\varepsilon, g);\ (\alpha_1, \beta_1), \dots, (\alpha_r, \beta_r)\},$$

subject to the conditions (i) $b \in Z$ *if* $\varepsilon = o_1$ & $b \in Z/_2$ *if* $\varepsilon = n_1$,

(ii) $0 < \beta_j < \alpha_j$, $(\alpha_j, \beta_j) = 1$ if $\varepsilon = o_1$, *and* $0 < \beta_j \leq \frac{\alpha_j}{2}$,

$(\alpha_j, \beta_j) = 1$ *if* $\varepsilon = n_1$.

Furthermore, if $\varepsilon = o_1$ *the equivariant diffeomorphism preserves orientation*.

Proof. It is not too hard given the discussion above to use a given family of invariants to build up an S^1-action. Conversely, given (M^3, S^1) classify M^* by (ε, g). Theorem 1.2

implies that we can find mutually disjoint tubular neighbourhoods of the finite family of exceptional orbits $E_1,\dots,E_r$. As above each of these can be described by the pair (α_j,β_j), subject to orientation conventions. The class b is now the obstruction associated to the bundle of principal orbits. Given two manifolds with matching invariants it is also clear that we can step-by-step construct an equivariant diffeomorphism between them.

Remark: when $\varepsilon = o_1$, $-M$ is specified by $\{-b-r;(o_1,g);(\alpha_1, \alpha_1-\beta_1)\dots(\alpha_r, \alpha_r-\beta_r)\}$.

Example: $M^3 = (D_1^{\,2} \times S_1^1) \cup_F (D_2^{\,2} \times S_2^1)$, where (a) F is the diffeomorphism of the common boundary defined by the matrix $\begin{pmatrix}-1 & 0\\ -m & 1\end{pmatrix}$, (b) S^1 acts trivially on $D_i^{\,2}$ and by rotation on $S_i^{\,1}$. Then F is equivariant (exercise) and the action has no exceptional orbits. The manifold M^3 coincides with the lens space $L^3(-m, 1) = L^3(m, m-1)$ and the orbit invariants are $\{-m;(o_1,0)\}$. The reader may find this clearer having read the section below on the fundamental group, when it will also emerge that the same space may correspond to more than one family of orbit invariants. Obviously two spaces may be diffeomorphic without being equivariantly diffeomorphic.

Motivated by the classification theorem 1.3 for pairs (M^3,S^1) we recall the original definition of a Seifert Manifold as a manifold M^3 which (a) decomposes into a collection of simple closed curves (called fibres), in such a way that each point belongs to one and only one fibre and (b) each fibre has a tubular neighbourhood N, consisting of fibres, such that $N = D^2 \times S^1/Z/\mu$ with the cyclic group Z/μ acting as in the discussion of exceptional orbits.

Such a manifold is the total space of a Seifert bundle - these are defined analogously to fibre bundles, except that the local product structure must be weakened to allow orbit spaces like N above. The coordinate transformations are given by maps $\gamma_{ij} ; \overset{o}{D}_i \cap \overset{o}{D}_j \to G$ as usual, which are compatible with the action of the finite "isotropy groups" $G_i \subseteq G$ on $\overset{o}{D}{}^2_i \times F$. A good reference for the general theory is [Ho]; in our case $F = S^1$ and $G = Top(S^1)$. The reduction of the structural group to some closed subgroup H of G is possible via the construction of a section of the associated Seifert bundle with fibre $G/_H$ - here, since the inclusion $i : O(2) \hookrightarrow Top(S^1)$ is well-known to be a homotopy equivalence, M^3 can be described as a bundle with O(2) as a structural group. Note that Seifert's condition (b) implies that each finite isotropy group G_i is already contained in O(2).

The classification theorem extends to cover this wider class of examples. The only new ingredient is that O(2) contains reflections of the generic fibre, hence along some non-trivial curve in M* the fibre may reverse its orientation. As in other contexts this phenomenon is described by a homomorphism $w : \pi_1 M^* \to Z/_2$. Among others we now obtain the subclass n_2 : M* is non-orientable, all generators of

$\pi_1 M^*$ reverse orientation and hence the total space M is orientable. Up to O(2)- diffeomorphism a manifold with $\varepsilon = n_2$ is specified by the orbit invariants

$$\{b\ ;\ (n_2,g);(\alpha^1,\beta_1)\dots(\alpha^r,\beta_r)\}\ ,\quad b \in Z,\ 0 < \beta_j < \alpha_j,(\alpha_j,\beta_j)=1.$$

It is important to understand this subclass, because it plays an important role in Chapters III and IV below.

Fundamental groups. In terms of the notation already introduced $\pi_1(M)$ has a presentation as follows:

Generators: $h,\ q_o,\ q_1\dots q_r \begin{cases} a_1\ b_1\dots a_g b_g & M^* \text{ orientable} \\ v_1\dots v_g & M^* \text{ non-orientable.} \end{cases}$

Relations:

$$\left.\begin{array}{l} a_i h a_i^{-1} = h \\ b_i h\ b_i^{-1} = h \end{array}\right\} (o_1)$$

$$v_i h v_i^{-1} = h^\eta \quad \begin{cases} \eta(n_1) = 1, \text{ M non-orientable} \\ \eta(n_2) = 1, \text{ M orientable} \end{cases},$$

$q_j h = h q_j$, $j = 1,\dots,r$.

(these are the commuting relations.)

$$q_j^{\alpha_j} = h^{\beta_j} = 1$$

(Geometrically the meridian m equals $\alpha q + \beta h$.)

$$q_o(q_1\dots q_r[a_1,b_1]\dots[a_g,b_g]) = 1 \text{ or}$$

$$q_o(q_1\dots q_r\ v_1^2\ v_1^2\dots v_g^2) = 1$$

$$q_o h^b = 1.$$

(the last two relations arise from the definitions of the section q_o and the obstruction b respectively).

Let C_u^A denote the finite cyclic group of order u generated by A. If $(s,u) = 1$, $1 \le s < u$, the representation ρ which maps A to

$$\begin{pmatrix} \zeta & o \\ o & \zeta^s \end{pmatrix}, \quad \zeta = e^{2\pi i/u},$$

induces a free action by C_u on S^3. Here free means that the isotropy subgroup of each point consists of the identity, and hence the orbit space S^3/ρ has the structure of a (smooth) 3-dimensional manifold, called the Lens Space, L(u,s). Since L(u,s) has S^3 as its universal covering space

$$\pi_1(L(u,s),\cdot) = C_u^A = H_1\ (L(u,s),\ Z).$$

Both S^3 and L(u,s) inherit an orientation from $\mathbb{C}^2$, with respect to which

$$L(u,s) = -L(u,u-s).$$

This follows by applying the orientation reversing differentiable involution

$$(z_1,z_2) \mapsto (z_1,\bar{z}_2).$$

LEMMA 1.4 <u>If the Seifert fibration</u> M^3 <u>has a finite cyclic fundamental group, then it has type</u> (o_1), $g = 0$ <u>and there at most two exceptional fibres.</u>

Proof. The universal covering space of M^3 must be homotopy equivalent to S^3, and each covering transformation orientation preserving. Since the fundamental group is abelian $((o_1),0)$ is a necessary condition. Suppose that $r > 3$, and consider

$$G_1 \underset{\langle h\rangle}{*} G_2 ,$$

where $G_i = \{q_i, h : q_i h q_i^{-1} = h,\ q_i^{\alpha_i} h^{\beta_i} = 1\}$, $i = 1,2,...r$.

The subgroup generated by h and $q_1 q_2$ is free abelian of rank 2. This is also the case for the subgroup generated by h and $q_3 \ldots q_r h^{-b}$ in

$$G_3 \underset{\langle h\rangle}{*} G_4 \underset{\langle h\rangle}{*} \ldots \underset{\langle h\rangle}{*} G_r$$

and in forming $\pi_1(M,\cdot)$ we may amalgamate along these subgroups. The generators $\{a_1, b_i\}$ or $\{v_i\}$ do not enter, since we assume that $g = 0$. It follows that $\langle h\rangle$ has infinite order, indeed is the unique maximal cyclic normal subgroup of G.

If $r = 3$ we distinguish between the cases $\frac{1}{\alpha_1} + \frac{1}{\alpha_2} + \frac{1}{\alpha_3} \leq 1$ and $\frac{1}{\alpha_1} + \frac{1}{\alpha_2} + \frac{1}{\alpha_3} > 1$.

For the latter the possibilities are $(2,2,\alpha_3)$, $(2,3,3)$, $(2,3,4)$ and $(2,3,5)$, which give rise to finite non-abelian fundamental groups, see the discussion of prism manifolds below. For the former the subgroup generated by h is normal, and the quotient group $G/_{<h>}$ (generated by $\bar{q}_i$ of order α_i, $i = 1,2,3$) is an infinite planar discontinuous group. Hence we are left with the cases $\{b; (o_1,0); (\alpha_i,\beta_i), r = 0,1,2\}$.

Before the next lemma we give a simple example of equivariant plumbing. A principal SO(2)-bundle over S^2 is classified by a map $S^1 \to S^1$ of degree $-m$; denote the associated D^2-bundle by Y_m, so that Y_{-1} is the disc bundle whose boundary S^3 has the Hopf fibration. More generally ∂Y_m is obtained as the union of two solid tori

$$\partial Y_m = B_1{}^2 \times S_1{}^1 \underset{F}{\cup} B_2{}^2 \times S_2{}^1 \quad ,$$

where F has the matrix

$$\begin{pmatrix} -1 & 0 \\ -m & 1 \end{pmatrix},$$

and it is well known that this identification gives the lens space $L(-m,1) = -L(m,m-1)$. We plumb Y_{m_1} to Y_{m_2} by choosing 2-discs B_1 and B_2 in the base spheres, and then identifying the covering copies of $D^2 \times D^2$ by means of the homeomorphism which interchanges the factors. Given the weighted graph

$$\overset{-b_1}{\bullet} \text{———} \overset{-b_2}{\bullet} \quad \cdots\cdots \quad -b_k$$

We plumb Y_{-b_i} to $Y_{-b_{i-1}}$ and $Y_{-b_{i+1}}$ using a pair of disjoint discs in S_i^2 $(i = 2,\ldots,k-1)$. Confining attention to the boundary of this 4-manifold (with any corners smoothed) we see that it is obtained by the identification of two solid tori by means of the product matrix

$$\begin{pmatrix} -1 & 0 \\ b_k & 1 \end{pmatrix} \begin{pmatrix} 0 & 1 \\ 1 & 0 \end{pmatrix} \begin{pmatrix} -1 & 0 \\ b_{k-1} & 1 \end{pmatrix} \cdots\cdots \begin{pmatrix} 0 & 1 \\ 1 & 0 \end{pmatrix} \begin{pmatrix} -1 & 0 \\ b_1 & 1 \end{pmatrix}.$$

The resulting 3-manifold carries an SO(2)-action such that all orbits are principal with the possible exceptions of the central curves of $B^2_{1,1} \times S^1_{1,1}$ and $B^2_{k,2} \times S^1_{k,2}$. Multiplying out the matrices we see that we have actually given an alternative description of the lens space

$L(u,s)$, where

$$\frac{u}{s} = b_1 - \cfrac{1}{b_2 - \cfrac{1}{\ddots \cfrac{}{\dfrac{1}{b_k}}}} = [b_1 \ldots b_k]$$

(continued fractions), see[O, page 26].

LEMMA 1.5 <u>If the Seifert fibration M^3 has a finite cyclic fundamental group</u>, M^3 <u>is a lens space. There are three possibilities</u>:

$r = 0$, $\{b;\ (o_1,\ 0)\} = L(-b,\ 1)$,

$r = 1$, $\{b;\ (o_1,\ 0);\ (\alpha,\beta)\} = L(b\alpha + \beta,\ \alpha)$

$r = 2$, $\{b;\ (o_1,\ 0);\ (\alpha_1,\ \beta_1),\ (\alpha_2,\ \beta_2)\} = L(b\alpha_1\alpha_2 + \alpha_1\beta_2 + \alpha_2\beta_1,$

$m\alpha_2 - n\beta_2)$,

with $m\alpha_1 - n(b\alpha_1 + \beta_1) = 1$.

Proof. In all three cases the order of the fundamental group follows from the table of generators and relations already given. For example, if $r = 1$, we have generators h and q_1 with relations $q_1^{\alpha} h^{\beta} = 1$, $q_1 h^{-b} = 1$, from which it follows that q_1 has order $\beta + b\alpha$. The integer s is rather harder to determine - if $r = 0$, and M^3 admits a principal $SO(2)$-action, M^3 fibres over $S^2 = CP(1)$ and the lens space must be symmetric, i.e. $s = 1$. If $r = 1$ or 2 we appeal to the alternative description of M^3 by means of an equivariant plumbing above. Inspection of the exceptional orbits, see [O, page 29] shows that the manifolds obtained from

(a) the orbit invariants $\{b;\ (o_1,0);\ (\alpha,\beta)\}$ and

(b) the weighted graph

$$\overset{-b-1}{\bullet}\!-\!\!-\!\!-\overset{-b_1}{\bullet}\quad\cdots\quad\overset{-b_k}{\bullet}\ ,\ \text{with } \frac{\alpha}{\alpha-\beta} = [b_1 \ldots b_k]\ ,$$

are equivariantly homeomorphic. We now have

$$\frac{u}{s} = [b+1, b_1,\ldots,b_k] = b+1 - \frac{1}{\frac{\alpha}{\alpha-\beta}} = \frac{\alpha(b+1)-(\alpha-\beta)}{\alpha} = \frac{b\alpha + \beta}{\alpha} .$$

The calculation for two exceptional orbits is similar, but more complicated, and is left as an exercise for the reader. However it is worth pointing out that in Chapter III below we never need the precise value of s; it is enough to know that a Seifert fibre space with cyclic fundamental group is a lens space, i.e. the orbit space of some free linear action by C_n on S^3 . We state this as

LEMMA 1.6 Let the cyclic group C_n act freely and topologically on S^3 . If the orbit manifold admits a Seifert fibration it is homeomorphic to a lens space.

Proof. Apply Lemma 1.4 and Lemma 1.5.

Now suppose the non-abelian group G admits a faithful representation in SO(4), inducing a free action on S^3. In this case the orbit space is called a prism manifold, and we will describe the groups G which arise, together with their representations in the next chapter. We shall also prove that up to homeomorphism a prism manifold is determined by its fundamental group, i.e. we avoid any problems associated with the integer s arising in the definition of the lens spaces. The Seifert manifolds with orbit data

$\{b;\ (o_1,0);\ (\alpha_i,\beta_i).\ i = 1,2,3,\ \sum_{i=1}^{3} \frac{1}{\alpha_i} > 1\}$
are prism manifolds; a less familiar description is given by

LEMMA 1.7 Let α be odd. The orbit invariants $\{b;\ (n_2,1);\ (\alpha,\beta)\}$ define a Seifert fibering of a prism manifold with fundamental group isomorphic to $C_\alpha \times D^*_{4|b\alpha+\beta|}$

(Here D^*_{4m} is the generalised quaternion group generated by $e^{\pi i/m}$ and j, discussed in detail in Chapter II below).

Proof. Using the table of generators and relations π_1 has a presentation

$$\{q,v,h:\ vhv^{-1} = h^{-1},\ qh = hq,\ q^\alpha h^\beta = 1,\ qv^2 = h^b\}.$$

The subgroup $\langle v^2,h,q\rangle$ is an abelian, indeed a cyclic, subgroup of index 2. Hence this subgroup is normal, and $v^4 = 1$. The remaining relations give

$$q^2 = h^{2b} \text{ and } q^{-\alpha} = h^\beta .$$

These equations combined with the assumption that α is odd imply that

$$h^{2|b\alpha+\beta|} = 1, \text{ so that}$$

h and v generate a subgroup of quaternion type of order $4|b\alpha+\beta|$.

The non-standard fibering of the prism manifold just given will be very important in the second part of Chapter III. It shows that such a manifold contains an embedded Klein bottle K^2 - look at the generators v and h.

Notes and references

Where it can be shown to exist a Seifert fibration has shown itself to be one of the most fruitful ways of studying a compact 3-manifold. Such fibrations were originally studied in the 1930's by H. Seifert and W. Threlfall, more up to date references are the lecture notes of P. Orlik [O] and Chapter VI in the book of J. Hempel [H]. I have mainly followed the former, emphasing those aspects which are important as prerequisites for Chapters III and IV below. I have also profited from the paper [O-W], in which the 3-dimensional theory is treated as a special case in the study of compact group actions on a smooth manifold of arbitrary dimension.

CHAPTER II: GROUPS WITH PERIODIC COHOMOLOGY

The aim of this chapter is to determine those finite groups which (a) have cohomological period dividing 4 and (b) contain no non-central elements of order 2. Both conditions are necessary for the existence of a free topological action on S^3, see [Mi 1]. We shall also determine those groups which are known to act freely and linearly on S^3; the second family of groups is properly contained in the first, and in dimensions $8k + 3$, $k \geq 1$, certain groups $Q(8n;\ k,\ell\ )$ defined below are known to act freely, but non-linearly.

Classically one knows that a finite subgroup of SO(3) is either cyclic, dihedral or one of three polyhedral groups (of orders 12, 24 and 60). Each of these determines a "binary" preimage in SU(2) the double covering group of SO(3). If q is a quaternion (real 4-vector on forgetting the multiplicative structure) and (a,b) is a pair of unit quaternions (elements of Sp(1) = SU(2)), then the map

$$q \mapsto aqb^{-1}$$

defines a two to one homomorphism of SU(2) × SU(2) onto SO(4). Given this homomorphism and the list of finite subgroups of SU(2) it is not hard to find all the finite subgroups of SO(4) and hence the members of the second family above. The only complications arise at the primes 2 and 3; this will become clear from the argument below.

If the finite group G acts freely and topologically on S^3 then by splicing together copies of the equivariant chain complex of S^3 we obtain a complex resolution for the cohomology of G, which obviously has period 4. This shows the necessity of (a) above. In terms of abstract group theory this implies that for all primes p dividing the order of G each subgroup of order p^2 is cyclic (p^2-condition). This in turn implies that a p-Sylowsubgroup of G is either cyclic or generalised quaternion, the latter class arising only for p = 2 The necessity of condition (b) implies that each subgroup of G of order 2p is cyclic (2p-condition). We consider first the case when all Sylow subgroups are cyclic - it is well-known, see for example [Ha] Theorem 9.4.3, that G is solvable and has a presentation

$G = \{A,B: A^m = B^n = 1, A^B = A^r, ((r-1)n, m) = 1, r^n \equiv 1 \pmod m\}$.

Such a group will be said to be of Type I. Let d equal the order of r in the group of units $Z/_m^x$. Type I groups satisfy the p^2-condition, and also the 2p-condition, provided that if d is even then so is $n/_d$.

LEMMA 2.1 _If_ G _is a Type_ I _group of cohomological period_ 4, _then_ d = 1 _or_ 2

Proof. The spectral sequence associated with the extension of C_m^A by C_n^B collapses, because m and n are coprime. The cohomological period thus depends on the fibre terms

$$E_2^{0,j} = H^j(C_m^A, Z)^B,$$

where the superscript B denotes the invariant elements. The relations $A^B = A^r$ and $r^d \equiv 1 \pmod m$ imply that the cohomological period exceeds 4 unless d = 1 or 2.

Remark: if d = 1, G is cyclic. Note also that AB^d generates a cyclic subgroup of G of order $\frac{mn}{d}$.

If X is an element of odd order in the subgroup generated by B, then the subgroup <X,A> is abelian, and hence cyclic. This follows from the equation $B^2AB^{-2} = A^{r^2} = A$. Hence G decomposes as a direct product.

$$G = C_{n_1}^X \times \langle A,\bar{B}\rangle,$$

where $n = n_1 2^t$, $\bar{B}$ has order 2^t, m is necessarily odd, and we denote the second factor by D'_{m2^t}. We have already noted that the 2p-condition implies that, if G is not cyclic G contains a subgroup of index two generated by AB^2. If α is a faithful one dimensional representation of $\langle AB^2 \rangle$ then the 2-dimensional induced representation $i_! \alpha$ defines a free linear action of G on $S^3 \subseteq \mathbb{C}^2$. So far as the generator A goes this is obvious; for the generator B note that B interchanges the two $\langle AB^2 \rangle$-cosets, and that B^2 acts freely inside the original one dimensional module. Alternatively the 2×2 matrix representing the action of B has neither eigenvalue equal to one. This proves half of

LEMMA 2.2 *If G is a finite group of Type I with cohomological period 4 satisfying the 2p-condition, then G acts freely and linearly on S^3. Furthermore if G is not cyclic then the orbit manifold $S^3/_G$ is unique up to homeomorphism.*

Proof. Since the irreducible complex representations of G have degree equal to 1 or 2 the only choice in the representation defining S^3/G is in the faithful one dimensional representation of $\langle AB^2 \rangle$. If ζ is a fixed primitive $(mn/_2)$th. root of unity, we may replace ζ by ζ^q, where $(q, mn) = 1$. However all such representations are interchanged by automorphisms of the form

$$AB^2 \longmapsto (AB^2)^q .$$

Note that the restricted representation $i^!_{G \to \langle AB^2 \rangle} i_{!\langle AB^2 \rangle \to G}(\alpha)$ has symmetric eigenvalues ζ^q, $\bar{\zeta}^q$. This uniqueness argument fails if G is cyclic, since a free linear action is determined by two independent one dimensional representations. Up to homeomorphism the orbit manifold (the lens space $L(m,q)$) is determined by the pair of eigenvalues (ζ, ζ^q), where $(q,m) = 1$, of the matrix representing A.

If $G = C_{n_1} \times D'_{4m}$, i.e. $t = 1$ in our earlier notation, the manifold is an example of a prism manifold, and the group D'_{4m} is a special type of generalised quaternion (or binary dihedral) group. With no restriction on m we denote these by

$$D^*_{4m} = \{A, B: A^{2m} = 1, A^m = B^2, A^B = A^{-1}\}.$$

If m is a power of 2, these groups arise as 2-Sylow subgroups for groups with periodic cohomology ($\mathcal{P}$-groups). In extending our discussion from Type I to more general groups it is convenient to work with the larger class of $\mathcal{P}'$-groups - closed both with respect to subgroups and homomorphic images - put another way we allow dihedral as well as cyclic and binary dihedral 2-groups.

In order to avoid an appeal to deep results in finite group theory we shall state the classification theorem only for solvable $\mathcal{P}$-groups, this was originally proved by Zassenhaus [Z].

Given groups N and Q any extension ($N \rightarrowtail E \twoheadrightarrow Q$) determines a homomorphism $h: Q \to \mathrm{Out}(N)$, the group of outer automorphisms of N. If Z is the centre of N there is a natural restriction map $r: \mathrm{Out}(N) \to \mathrm{Aut}(Z)$, and given (N,Q,h) we may regard Z as a Q-module via $r \circ h$. There exist extensions corresponding to this data if and only if a certain obstruction in $H^3(Q,Z)$ vanishes, and these are classified as elements in $H^2(Q,Z)$. Note that both groups vanish if either $Z = \{1\}$ or if Q and Z have coprime orders. Fundamental to our discussion are the groups

$SL(2,\mathbb{F}_3) = T^*_1 = \{X, D^*_8: X^3 = 1, A^X = B, B^X = AB\}$, a split extension of D^*_8 by C_3, and the quotient group $PSL(2,\mathbb{F}_3) = T_1$, obtained by dividing out by the centre. T_1 and T^*_1 label these groups as the tetrahedral and binary tetrahedral group respectively, and we also have the pairs (O_1, O^*_1), (I, I^*) associated with the other platonic solids. However note that I and I* are not solvable; we shall return to them in a later chapter. Both T_1 and T^*_1 have outer automorphism groups of order 2.

LEMMA 2.3 (i) Any extension of T_1 or T^*_1 by a group of odd order splits as a direct product.

(ii) The only extension of T_1 by C_2 which is a $\mathfrak{P}'$-group is $PGL(2,\mathbb{F}_3)$

(iii) The only extension of T^*_1 by C_2 which is a $\mathfrak{P}'$-group is O^*_1 .

Proof. (i) is immediate, since the outer automorphism group has order 2, and the centre is either trivial (T_1) or has order $2(T^*_1)$.

(ii) Since Z is trivial the extensions are distinguished by Hom $(C_2, \mathrm{Out}(T_1))$ of order 2. One of these is $PGL(2,\mathbb{F}_3)$, the other does not belong to the class $\mathfrak{P}'$.

(iii) This is similar to (ii), except that we must calculate $H^2(C_2, Z)$ of order 2. There are four possible extensions, those corresponding to the trivial map h have 2-Sylow subgroup $C_2 \times D^*_8$ (not in $\mathfrak{P}'$), the remaining two have 2-Sylow subgroup semidihedral (not in $\mathfrak{P}'$) or D^*_{16} (corresponding to O^*_1). We note that O^*_1 has a presentation $\{R, T^*_1 : \langle R, D^*_8\rangle = D^*_{16}, X^R = X^{-1}\}$.

In the proof below we shall refer to the groups $SL(2,\mathbb{F}_3)$, $PSL(2,\mathbb{F}_3)$, $PGL(2,\mathbb{F}_3)$ and O^*_1 as exceptional.

THEOREM 2.4 Let G be a solvable $\mathfrak{P}'$-group, and let A_G be a maximal normal subgroup of odd order. Then $G/_{A_G}$ is either a 2-group or exceptional.

Proof. Inductively we suppose that G has a normal subgroup H such that H satisfies the conclusion of the theorem and $G/_H$ is cyclic of prime order. A_H is a characteristic subgroup of H, hence normal in G, so we may factor out by A_H and without loss of generality assume that H is either a 2-group or exceptional.

Case (i), H is exceptional. By Lemma 2.3 G is isomorphic to $H \times C_p$ (p=odd) or $G = PGL(2,\mathbb{F}_3)$ or $O^*_1 (p = 2)$.

Case (ii), H is a 2-group (cyclic, quaternion or dihedral). We need only consider the case when $G/_H$ has odd prime order, since if $G/_H = C_2$, G will have 2-Sylow subgroups outside the class $\mathcal{P}'$. Hence it is enough to classify (split) extensions of C_{2^t}, $D^*_{2^t}$ and D_{2^t} by C_p. The splitting exists because the orders are coprime.

Since $Aut(C_{2^t})$ has order 2^{t-1} the cyclic case is trivial, for the two others note that Aut(H) only contains elements of odd order if $H = C_2 \times C_2$ or D^*_8. In both cases $p = 3$, and the non-trivial extensions are $PSL(2,\mathbb{F}_3)$ and $SL(2,\mathbb{F}_3)$ respectively.

For $\mathcal{P}$-groups as as opposed to $\mathcal{P}'$-groups we have proved that $G/_{A_G}$ is isomorphic to one of the groups C_{2^t}, $D^*_{s^t}$, T^*_1 or O^*_1, but before giving the classification in its most useful form we must allow for the possibility that 3 divides the order of A_G in the last two cases. Thus we must describe all extensions of $C_{3^{v-1}}$ by either T^*_1 or O^*_1 which satisfy the $\mathcal{P}$-condition.

Consider $(C_{3^{v-1}} \rightarrowtail E \twoheadrightarrow T^*_1)$. Because of coprime orders the induced extension E' of $C_{3^{v-1}}$ by D^*_8 is trivial, provided that the module structure given by $h: D^*_8 \to Aut(C_{3^{v-1}})$ is trivial. But this must be the case, since the D^*_8-action is the lift of a T^*_1-action, which factors through the abelianised group and hence maps D^*_8 to the identity. Therefore $E' \cong C_{3^{v-1}} \times D^*_8$, D^*_8 is normal in the required extension E, the quotient is cyclic on order 3^v and there is only one possible action. This unique extension E is usually denoted by $T^*_{3^v}$ with presentation
$\{X, D^*_8: X^{3^v} = 1, A^X = B, B^X = AB\}$.

Finally consider $(C_{3^{v-1}} \rightarrowtail E'' \twoheadrightarrow O^*_1)$, which has D^*_{16} as 2-Sylow subgroup, and which contains the extension E just constructed as a (normal) subgroup of index 2. The extension E" is determined by an element in

$H^2(C_2, Z(T^*_{3^v}))$ (of order 2) and by the homomorphism $h: C_2 \to Out(T^*_{3^v})$. If we note that any automorphism of $T^*_{3^v}$ induces automorphisms of the central subgroup $C_{3^{v-1}}$ and of the quotient T^*_1, it is clear that E" is determined by the same data as determines $(D^*_8 \rightarrowtail D^*_{16} \twoheadrightarrow C_2)$. Hence E" is again unique and is usually denoted by $O^*_{3^v}$. We shall not give a presentation of $O^*_{3^v}$ for $v \geq 2$, since it plays virtually no part in our later arguments.

We can summarise the classification achieved so far by

THEOREM 2.5 _If_ G _is a solvable_ $\mathcal{P}$_-group, then_ G _is a split extension of a Type_ I _group_ K _by_ C_{2^t}, $D^*_{2^t}$, T^*_v _or_ O^*_v. _In the first case_ G _is again of Type_ I, _and we label the remaining cases Types_ II, III _and_ IV. _The order of_ K _is odd for Type_ II _and coprime with_ 6 _in the remaining cases._

Next consider the conditions under which the cohomological period equals 4, i.e. the cohomological period for each prime dividing the order equals 2 or 4. Using calculations of R.G. Swan [Sw 1] the subgroup isomorphic to $G/_K$ defining the type has cohomological period 2 (Type I) or 4 (Types II, III, IV). (For all groups except O^*_v $(v \geq 2)$ this is clear from the representation theory below.) Thus the problem is reduced to finding the cohomological period of K; from the discussion in Lemma 2.1 the structural constant $d = 1$ or 2. However $d = 2$ is impossible, since d divides n the order of the generator B, which must be odd. Hence $d = r = 1$ and the subgroup K is cyclic, generated by $Y = AB$.

For Type III we can say more, since the action of T^*_v on $\langle Y \rangle$ is determined by the homomorphism h of T^*_v into an abelian group, hence h is trivial on the commutator subgroup D^*_8 (compare the construction of T^*_v from T^*_1 above). But h must also be trivial at the prime 3, since otherwise G would contain a metacyclic group of order $3^v mn$, which would necessarily have period exceeding 4. Therefore a Type III group of cohomological period equal to 4 must be isomorphic to a direct product,

$$\langle Y \rangle \times T^*_v .$$

Unfortunately the situation is not so simple for Types II and IV, since the commutator quotients of $D^*_{2^t}$ and O^*_v contain 2-torsion. Consider the most general extension

$$\langle Y \rangle \rightarrowtail G \twoheadrightarrow D^*_{2^t} .$$

The quotient acts on $K = \langle Y \rangle$ as a pair of commutating involutions, each of which splits K as $K^+ \times K^-$, where K^+ consists of elements which are inverted. The second involution introduces further splittings which we may denote by

$$K = K^{++} \times K^{+-} \times K^{-+} \times K^{--} ,$$

and it follows that G has a presentation $\{Y_1, Y_2, Y_3, Y_4, A, B:\ Y_i^{m_i} = 1,\ \langle A,B \rangle = D^*_{2^t},\ [Y_1,A] = [Y_1,B] = [Y_2,A] = [Y_3,B] = 1,\ Y_2^B = Y_2^{-1},\ Y_3^A = Y_3^{-1},\ Y_4^A = Y_4^B = Y^{-1}\}$, where the m_i are mutually coprime and odd, and the groups are denoted in [Mi 1] by $C_{m_1} \times Q(2^t m_2, m_3, m_4)$. We distinguish three subcases: $II\kappa$ ($m_3 = m_4 = 1$), $II\lambda$($t \geq 4$ and $m_3 m_4 \neq 1$) and $II\mu$ ($t = 3$, and at least two of the integers m_2, m_3 and m_4 are greater than 1).

Passing to Types III and IV, we have already seen that the former contains no subgroups of Types $II\lambda$ or $II\mu$. In Type IV, if $v \geq 2$, the model group O^*_v contains $Q(16, 3^{v-1}, 1)$ as a subgroup of index 3; more careful analysis shows that there can be no subgroups of Type $II\mu$. Hence if we assume that G is a solvable group which (a) has cohomological period dividing 4 (b) satisfies the 2p condition and (c) contains no subgroup of Type $II\lambda$ or $II\mu$, we have shown that G is a direct product of a cyclic group of coprime order with one of the groups

$$D'_{(2m+1)2^t}\ ,\ D^*_{4m}\ ,\ T^*_v (v \geq 1) \text{ or } O^*_1\ .$$

THEOREM 2.6 <u>If the solvable group</u> G <u>satisfies assumptions</u> (a), (b) <u>and</u> (c) <u>above, then</u> G <u>acts freely and linearly on</u> S^3 .

Proof. For Type I this is Lemma 2.2. The proof for Type II is similar, since G again contains a cyclic subgroup of index 2 and we can transfer a faithful one dimensional representation of this subgroup. Inspection of the eigenvalues for the matrix representing the generator B of D^*_{4m} again shows that the induced action on S^3 is free. Next consider T^*_1 of order 24: it is an easy exercise in representation theory to show that the degrees of the irreducible representations are 1(3 times) 3(once) and 2 (3 times), and that the first four representations factor through T_1 . If ω is a primitive cube root of unity, then the eigenvalues of the matrices representing the generator X in the remaining three representations are $(1,\omega)$, $(1,\omega^2)$ and (ω,ω^2). Only the last τ corresponds to a free action of T^*_1 on S^3. For O^*_1 the argument is similar (and helped by the fact that T^*_1 is a subgroup of index 2); there are two irreducible two dimensional complex representations $O_{\pm}$ with no eigenvalue equal to 1. These may be distinguished by their restrictions to D^*_{16}, for which we obtain the two representations which restrict in their turn to the standard inclusion of $D^*_8 = \{\pm 1, \pm i, \pm j, \pm k\}$ in $Sp(1) = SU(2)$. The group T^*_v $(v \geq 2)$ is a little more interesting - if we use the isomorphism of Spin(4) with $SU(2) \times SU(2)$ mentioned at the start of this chapter, and recall that T^*_v contains an index 3 subgroup isomorphic to $C_{3^{v-1}} \times D^*_8$, we see that any monomorphism into SO(4) must map the 3-Sylow subgroup generated by X to $(\zeta^{k+3^{v-1}} \oplus \zeta^{-k+3^{v-1}})$, where ζ is a primitive 3^v-th root of unity and $(k,3^v) = 1$. On the 2-Sylow subgroup D^*_8 the map is uniquely defined, and the numbers are so chosen that after projection onto one of the factors SU(2), which introduces a kernel generated by X^3, we obtain the unique fixed pointfree representation τ of T^*_1 described above.

In all cases we extend the action to a direct product with the cyclic group C_u by taking the tensor product of one of the representations just described with a faithful representation of C_u in $U(1)$.

Remark. There is a more detailed discussion of the representations of the binary polyhedral groups in [Th 1], we see also the book by J. Wolf [Wo].

We conclude this chapter with some remarks about the groups $Q(2^t m_2, m_3, m_4)$. If $t \geq 4$ an argument of R. Lee [L] shows that there is a non-vanishing surgery obstruction to the existence of a free (topological) action on S^{8k+3}. Hence if G is of type IV and G acts freely on S^3, there is also a free linear action. If G belongs to the remaining subclass IIμ, $G = Q(8m_2, m_3, m_4)$ and contains a subgroup $Q(8p, q)$, where p and q are distinct odd primes. It is easy to list the irreducible complex representations of this subgroup, using [Se, §9.2]. These either factor through one of the quotient groups D^*_{8p} or D^*_{8q}, or are induced up from the cyclic subgroup $C_{pq} \times \text{centre}(D^*_8)$ of order 2pq. Thus the possible dimensions over $\mathbb{C}$ are 1, 2 or 4 - note that

$$8p + 8q - 8 + 4^2\left(\frac{(p-1)(q-1)}{2}\right) = 8pq.$$

(the negative term -8 implies that we only count D^*_8 representations once, and the final term on the left hand side includes those representations which induce free linear actions on S^7.) We have already seen that $Q(8p,q)$ has cohomological period 4; for suitably chosen values of p and q it is possible to show that Q acts freely on S^{8k+3}, $k \geq 1$. In dimension 3 this means that there is no obstruction to finding a finite Poincare complex with fundamental group isomorphic to Q and universal cover homotopy equivalent to S^3. Since the surgery obstruction also vanishes we can even find a free action by Q on a homology 3-sphere, whether this can be taken simply connected should be regarded as an open question. The argument outlined in [Th 2] has a gap, the filling of which depends either on extending work of H. Miller to describe the unstable homotopy set [BQ, BSO(4)] or on an argument with KSp-valued characteristic classes to exclude the stable classes defined by certain virtual representations from this set. I have yet to complete the latter.

Notes and references.

The algebraic classification of solvable $\mathcal{P}'$-groups is due to H. Zassenhaus [Z]. Our treatment is essentially the same as his, except that we use group cohomology to solve the extension problems. For more details and for the extension of the classification to non-solvable groups, see the unpublished preprint [Th-W]. If G is a non-solvable $\mathcal{P}$-group, then at worst G/A_G contains a subgroup of index 2 isomorphic to $SL(2,\mathbb{F}_p)$, $p \geq 5$. Since the latter group has cohomological period equal to the lowest common multiple of 4 and (p-1), in considering free actions on S^3 the only new family of groups is $C_u \times SL(2,\mathbb{F}_5)$, $(u,30) = 1$.

The best reference for the geometric classification of free linear actions by finite groups on S^{2n-1} is the book of J. Wolf, *Spaces of constant curvature* [Wo], particularly Chapters 4 - 7. This includes a thorough description of the representations which can arise for the more complicated Types III and IV.

CHAPTER III: Free C_2 and C_3 actions on certain Seifert manifolds.

In this Chapter we shall present proofs of the theorems of R. Myers My , generalising earlier work of R. Livesay Li , and of J. Rubinstein R . Taken together these come close to a topological classification of free actions by groups of order $3^s 2^t$ on S^3 , and provide the major ingredient in the reduction theorem of the next chapter. As before let $L^3(m,q)$ be a lens space with cyclic fundamental group of order m.

THEOREM 3.1 *Let* h *be a free involution on* $M = L^3(m,q)$. *The orbit manifold* $M^* = M/h$ *is Seifert fibered, and hence homeomorphic to a manifold of constant positive curvature*.

The proof will be broken up into several lemmas.

LEMMA A *There is a* $p\ell$ *Morse function* $f:M \to R$ *and a triangulation* K *of* M, *such that*

(1) f *has* 4 *critical points* $\{x_o, x_1, x_2, x_3\}$ *such that* $f(x_i) = i = \text{index}\ (x_i)$,

(2) $A = \{x \in M: fh(x) = f(x)\}$ *is a closed orientable surface containing no critical points*,

(3) <u>f is linear on the simplexes of</u> K,

(4) h <u>is simplicial with respect to</u> K, <u>and</u>

(5) A <u>and</u> $\{x_i\}$, $0 \le i \le 3$, <u>are triangulated as full subcomplexes of</u> K.

Proof. The description of L(m,q) as a Seifert manifold with one exceptional fibre in the first section shows that M has a handle decomposition with one handle each of index i, $0 \le i \le 3$. The construction of the "height" function f satisfying (1), (3) and (4) is now easy. Furthermore we can arrange for the vertices of f to have distinct images under f. Let $M_1 = \text{graph}(f) \subseteq M \times R$, and $M_2 = (h \times 1)(M_1)$. Write $g: M \times R \to M$ for projection onto the first factor. If τ is an arbitrary 3-simplex in M_1, $(g\tau \times R) \cap (M_1 \cup M_2)$ consists of a union $\tau \cup \tau'$ in general position. Thus $\tau \cap \tau'$ contains no vertices and arguing with each 3-simplex in turn we find that $A = g(M_1 \cap M_2)$ is a closed two-sided sub-2-manifold, containing no critical points. This is (2) and (5) follows by taking a suitable h-equivariant subdivision.

LEMMA B $H_2(A^*, Z/_2) \to H_2(M^*, Z/_2)$ <u>is non-trivial</u>.

Proof. Let $x \in M$ and let α be a path such that $\alpha(0) = x$, $\alpha(1) = hx$.
Define $F:[0,1] \to R^1$ by $F(t) = f\alpha(t) - fh\alpha(t)$.
$F(0) = -F(1)$, so there exists an intermediate point t_o, $0 < t_o < 1$, such that $F(t_o) = 0$. Therefore $\alpha(t_o) \in A$, and every non-bounding symmetric 1-cycle in the dual subdivision

to K meets A. Taking the Poincaré dual, no symmetric 2-cocycle belongs to the image of $H^2_{Z/2}(M,A;Z/_2)$ in $H^2_{Z/2}(M;Z/_2)$. Dividing out by $Z/_2$ we see that $H^2(M^*,A^*;Z/_2) \to H^2(M^*,Z/_2)$ is trivial, so that $H^2(M^*,Z/_2) \to H^*(A^*,Z/_2)$ must be a monomorphism. The conclusion of the lemma is the dual of this statement.

LEMMA C <u>There is an element</u> $u \in H_1(A^*,Z)$ <u>of order</u> 2 <u>which has non-trivial image in</u> $H_1(M^*,Z)$.

Proof. This is an elementary consequence of Lemma B.

Denote the images under f of the K vertices by $y_o, y_2, \ldots, y_{2n}$, and set $y_{2_{j-1}} = \dfrac{y_{2_{j-2}} + y_{2_j}}{2}$, $1 \le j \le n$. Introduce the further notation

$$B_j = f^{-1}(y_j) \cap A,$$

$$X_j = f^{-1}([y_{2j-2}, y_{2j-1}] \cap A), \quad Y_j = f^{-1}([y_{2j-1}, y_{2j}] \cap A),$$

$$X = \bigcup_{j=1}^{n} X_j \cup B_{2n}, \quad Y = B_o \cup \bigcup_{j=1}^{n} Y_j, \quad B = \bigcup_{j=0}^{2n} B_j, \quad D = \bigcup_{j=0}^{n} B_{2j}.$$

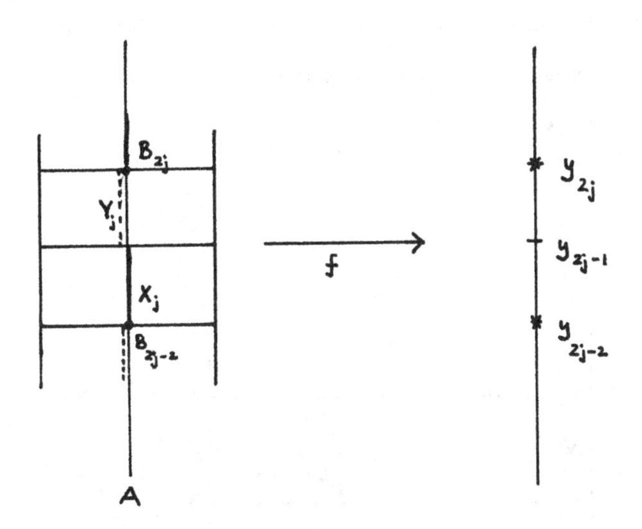

LEMMA D <u>There are deformation retractions of</u> X* (<u>and</u> Y*) <u>onto</u> D* .

Proof. We construct an equivariant retraction of X_j onto B_{2j-2}. The argument for Y_j is similar. Let $\sigma = \langle v_o v_1 v_2 \rangle$ be a typical simplex of A such that $\sigma \cap X_j \neq \emptyset$.

There are three possibilities:-

(i) f maps each vertex to y_{2j-2}, when there is nothing to prove.

(ii) $Fv_o \leq y_{2j-2}$, $fv_1 \leq y_{2j-2}$, but $fv_2 \geq y_{2j}$. Let $x \in \sigma \cap X_j$ and extend the line joining v_2 to x to meet $\langle v_o v_1 \rangle$. If $f(x) = sy_{2j-1} + (1-s)y_{2j-2}$, then with the notation as in the diagram below (to be thought of as being normal to the page),

$$x = s\alpha(x) + (1-s)\ (x),\ 0 \leq s \leq 1.$$

Recall that f is linear on simplexes from Lemma A. The points $\alpha(x)$ and $\beta(x)$ are the intersection points of $\langle v_2 x \rangle$ with B_{2j-1} and B_{2j-2} respectively.

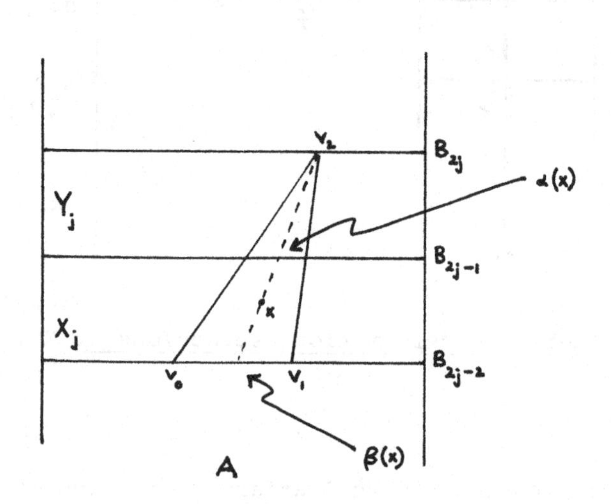

Set $H_j(x,t) = s(1-t)\alpha(x) + (1-s(1-t))\beta(x)$ for each $x \in \sigma \cap X_j$.

(iii) $f(v_0) \leq y_{2j-2}$, $f(v_1) \geq y_{2j}$, $f(v_2) \geq y_{2j}$.

The argument is similar to (ii), except that one needs a little care when $f(v_0) = y_{2j-2}$. In this case if $x = v_0$ set $H_j(x,t) = x$.

The retraction H_j behaves well on $\sigma_1 \cap \sigma_2$ and is equivariant.

Extending the usual definition we consider a knot γ to be an embedding of S^1 in M, and similarly for a link $\gamma_0 \cup \gamma_1$. The knot γ is trivial if it spans an embedded D^2, and toral if it lies on a torus T^2 splitting M into the union of two solid tori. A toral link is similarly defined.

LEMMA E (M,h) contains either (i) a trivial knot γ such that $h\gamma = \gamma$, or (ii) a non-trivial toral knot γ such that $h\gamma = \gamma$ or (ii) a non-trivial torus link $\gamma_o \cup \gamma_1$ such that $h\gamma_i = \gamma_{i+1 (\text{mod } 2)}$.

Proof. This starts with a diagram chase

$$(x,y) \longmapsto u \underset{}{\longmapsto} 0 \quad \text{(}u\text{ of order 2)}$$

$$w \overset{i_1}{\longmapsto} x$$

$$\to H_1B^* \xrightarrow[(i_1,i_2)]{} H_1X^* \oplus H_1Y^* \xrightarrow[(j_1-j_2)]{} H_1A^* \xrightarrow{\partial} H_0B^* \text{ (free abelian)}$$

$$H_1X^* \oplus H_1Y^* \cong H,D^*$$

Set $z = i_2w-y$.

Then $j_2 = (j_1-j_2)((x,y)-(x,i_2w)) = (j_1-j_2)((x,y)) = u$.

Theorem $(j_1-j_2)(x,i_2w) = (j_1-j_2)(i_1w,i_2w) = 0$ by exactness.

Therefore by Lemma C there exists some component c^* of B_2^* such that the induced map $H_1C^* \to H_1M^*$ is non-trivial.

Geometrically this implies that there exists some simple closed curve $\gamma^* \in C^*$ such that $[\gamma^*] \neq 1$ in π_1M^*. This is the critical step in the proof.

The generic case is the following:

$$q^{-1}\gamma^* \subseteq B_{2_j} = f^{-1}(y_{2_j}) = \begin{cases} \text{a torus splitting } M \text{ if } 1 < y_{2_j} < 2 \\ S^2 \text{ if } 0 < y_{2_j} < 1 \text{ or } 2 < y_{2_j} < 3. \end{cases}$$

Subcase (α): $[\gamma^*] \notin$ Image $(\pi_1 M)$, and γ^* lifts back to an invariant simple closed curve, which is either trivial or toral.

Subcase (β): $[\gamma^*] \in$ Image $(\pi_1 M)$, and γ^* lifts back to a toral link.

The non-generic case arises when $y_{2_j} = 1$ or 2, when we have to allow for a pinched torus containing a critical point. Now exploit the assertion in Lemma A that A contains no critical points to show that the lift of γ^* avoids critical points in all cases.

Having found an invariant curve or pair of curves in M we conclude the proof of Theorem 3.1 quite easily. First we consider a non-trivial torus knot γ. As a simple closed curve γ is associated with two tori (i) the splitting torus $T^2 = f^{-1}(y_{2_j})$ and (ii) N the boundary of an equivariant tubular neighbourhood.

If γ is isotopic to the core of either "handle" in M, then M* inherits a similar splitting from M and is a lens space. Otherwise write $\overline{M-N} = V_o \underset{F}{\cup} V_1$, where F identifies part of the boundaries of two solid tori along an annulus $(\overline{T-N \cap T})$. The central loop in F is homologous to multiples $m_i \geq 2$ of the cores of the V_i $(i = 1, o)$, so that $\overline{M-N}$ is

Seifert fibered with exceptional fibres of multiplities m_0 and m_1 over an orbit space D^2. It follows that M^*-N^* is also Seifert fibered. (This step is not trivial. The arguments in Section 1 show that $\pi_1(M-N)$ contains a central cyclic summand, which since h preserves orientation survives into $\pi_1(\overline{M^*-N^*})$. As a manifold with boundary $\overline{M^*-N^*}$ is sufficiently large, and is Seifert fibred by an argument of F. Waldhausen, Topology 6(1967). See also the Chapter already cited in [H]) There are two possibilities on sewing back the solid torus N^*: (α) a meridian curve is not isotopic to a generic fibre, in which case we may extend the fibering of $\overline{M^*-N^*}$ to M^* with the core of N as an additional exceptional fibre. Otherwise (β) we have an "improper" Seifert fibration with fibre of type $(0, \pm 1)$. Looking at the presentation of π_1 given in section one we see that the generic fibre h has become null-homotopic, which leads to a lens space. Note that possibility (α) corresponds to the situation when $\pi_1(M^*)$ is non-abelian, that is, h acts non-trivially on the generator of $\pi_1(M) = C_m$.

Next suppose that M contains a toral link $\gamma_0 \cup \gamma_1$, where the two components are interchanged by the involution h. We now have three tori to consider, T (carrying $\gamma_0 \cup \gamma_1$) which defines a genus one Heegard splitting of M, and $N_i (i = 0,1)$ the boundaries of tubular neighbourhoods of the components of γ, also interchanged by h. Consider the annuli $N_0 \cap T, N_1 \cap T$; $\overline{M-(N_0 \cup N_1)}$ is the union of two solid tori $V_0 \underset{F_0 \cup F_1}{\cup} V_1$, where

$F_o \cup F_1 = \overline{T - T \cap (N_o \cup N_1)}$, and the core of F_i is homologous to a multiple m_i of the core of V_i, $i = 1,0$. We may suppose that $m_o \geq 2$, since if $m_o = m_1 = 1$, M is homeomorphic to $S^2 \times S^1$. It follows that $\overline{M - (N_o \cup N_1)}$ is Seifert fibered with one or two singular fibres of multiplicities m_o, m_1, and the orbit surface is an annulus. The argument concludes in the same way as the toral knot case.

Finally the case when γ is a simple closed curve is that considered by R. Livesay, the manifold M is S^3 and the involution h is conjugate to the antipodal map.

Our next aim is to at least indicate the method of proof of three related results of J. Rubinstein [R2]. Each of these concerns free actions by C_3 on a manifold M^3 belonging to the class *M* of closed, orientable, irreducible 3-manifolds having π_1 finite and an embedded projective plane RP^2 or Klein bottle K^2. We give full details of the first result, since this contains the essential ideas, but then content ourselves with an outline of the other two.

THEOREM 3.2 <u>A free topological action by C_6 on S^3 is conjugate to a free linear action.</u>

Proof. By Theorem 3.1 it is enough to consider a free C_3^A action on RP^3. Choose $P = RP^2$ in equivariant general position, so that each of $P \cap AP$, $P \cap A^2P$ and $AP \cap A^2P$ consists of a family of simple closed curves, intersecting only in isolated triple points. There is a unique component of $P \cap AP$ which is 1-sided in P and all other curves are contractible. The argument is one of successive simplification of the intersection curves, by first removing all contractible curves, and then all but one last triple of triple points, control being exercised by counting the number of triples remaining after each spherical modification.

Removal of the contractible curves in $P \cap AP$. Let C be a curve in the intersection bounding a disc D in P with the interior $\mathring{D}$ disjoint from AP. If $D \cap A^2P = \emptyset$, and C bounds D_o in AP, we can reduce the number of intersection curves without introducing new triple points by replacing P by

$$P_o = (P - A^2D_o^c) \cup A^2D$$

(moved slightly away from A^2D). If we are not in this simple situation we must choose the disc for the modification more carefully.

Assume $D \cap A^2P \neq \phi$, and contains no loops. Let λ be an arc of $D \cap A^2P$ chosen so that there is an arc μ of C with $\partial\mu = \partial\lambda$ and $A^2P \cap \mathring{\mu} = \phi$, see Figure 1 below. Let D_1 be the smaller 2-disc contained in D bounded by $\lambda \cup \mu$. Then if $D_1 \cap AD_1 \neq \phi$, $\partial\lambda = \{x,Ax\}$ and

$$D_1 \cap AD_1 = \{Ax\},$$

(this follows since $\mathring{D}_1$ cannot meet $AP \cup A^2P$, and if $A\lambda$ were to meet μ in an interior point we would have $A\lambda = \mu$, forcing A to fix the end points, a contradiction.)

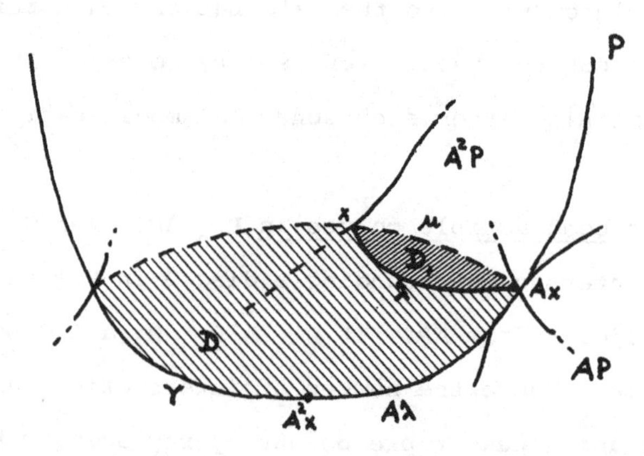

Fig. 1.

We may suppose that μ and $A\lambda$ are adjacent arcs in C and that λ is contained in A^2C. Furthermore $C \cap A^2C$ contains the triple points x, Ax and A^2x. Let

$$C' = \mu \cup A\lambda \cup A^2\mu,$$

a simple closed curve, which is again contractible because it does meet the non-contractible component of $P \cap AP$. By a slight A^2C at an odd number of points, another contradiction.

Given that $D_1 \cap AD_1 = D_1 \cap A^2D_1 = \emptyset$, let N be a small regular neighbourhood of D_1 with $D_2 = A^2P \cap N$. If D_3 is the disc in the boundary of N having the same one-dimensional boundary as D_2 (see Figure 2), then $D_2 \cup D_3$ bounds a 3-cell contained in N containing D_1. Then the arc $A\lambda$ may be moved across AD_1 past the arc $A\mu$ by means of an isotopy taking AD_2 to AD_3 across the 3-cell. This simplifies the intersection of the original disc D with A^2P, and reduces the number of triple points. After a finite number of subsidiary modifications of this kind, we may carry out the first step and remove C itself.

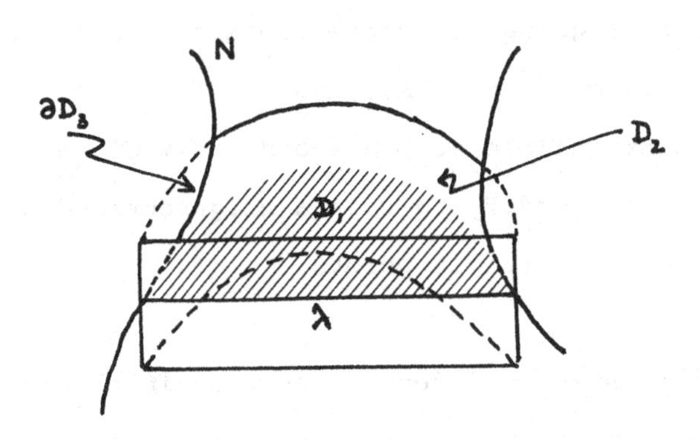

Fig. 2.

<u>Removal of triple points</u>. $P \cap AP = C$(non-contractible), $C \cap A^2C$ contains 3n triple points, where $n \geq 1$.

The first part of the argument can be applied to remove triple points, unless we are in the exceptional situation

$$\partial\lambda = \{x,Ax\},\ C' = \mu \cup A\lambda \cup A^2\mu,$$

see Figure 1 again. If C' is contractible in P we can argue as before to reach a contradiction, so assume that C' is non-contractible and that $n \geq 1$. In this situation we can find arcs $\lambda' \in C$, $\mu' \in A^2C$ with common end points, bounding a disc D' contained in P, the interior of which avoids both AP and A^2P. ($A^2C = P \cap A^2P$ and $C \cap A^2C$ contains only isolated points.) The disc D' actually avoids its translate AD', because otherwise, arguing as before the intersection would have to take the form $\partial\lambda' = \{y,Ay\}$, and $C'' = \lambda' \cup A\mu' \cup A^2\lambda'$ would be a non-contractible loop disjoint from C', a contradiction. Hence modifying P in a neighbourhood of AD' we reduce the number of triple points.

<u>Fibering the orbit space</u>. Referring once again to Figure 1 let ν be the complement in C to the arc containing the point Ax, with end points $\{x,A^2x\}$. Since C' is non-contractible the union of ν and the third arc in C', $A^2\mu$, bounds a disc D_4 in P which is such that

$$(AP \cup A^2P) \cap \mathring{D}_4 = \phi.$$

Similarly using the non-contractibility of A^2C we can find D_5 such that $\partial D_5 = A\lambda \cup A^2\nu$, and

$$(AP \cap A^2P) \cap \mathring{D}_5 = \emptyset$$

(observe that A^2C is bounded by $A^2\mu \cup \lambda \cup A^2\nu$)
Now $\partial D_1 = \lambda \cup \mu$, $\partial(AD_4) = \mu \cup A\nu$ and $\partial(A^2D_5) = \lambda \cup A\nu$, so that the union of D_1, AD_4 and A^2D_5 forms a 2-sphere bounding a 3-cell E. By applying A to the boundary of E we see that we may take E to satisfy $E \cap AE = \{Ax\}$; let α be a properly embedded unknotted arc in E with endpoints $\{x, Ax\}$. Let D_6, D_7, D_8 be discs embedded in E, which are such that (see Figure 3)

(i) the interiors are mutually disjoint and avoid the bounding sphere S, and

(ii) $\partial D_6 = \alpha \cup \mu$, $\partial D_7 = \quad \cup \lambda$, $\partial D_8 = \alpha \cup A\nu$.

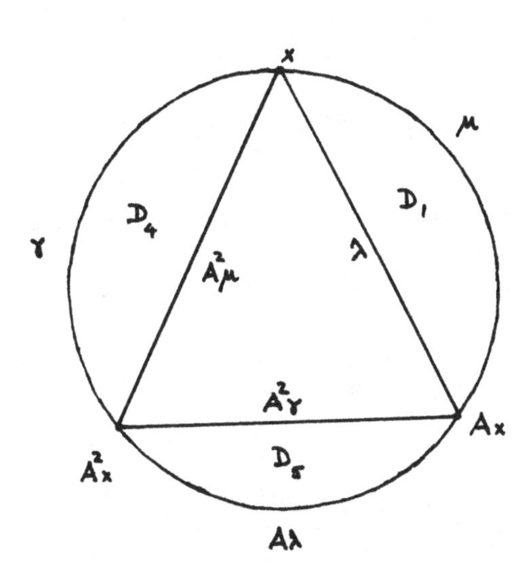

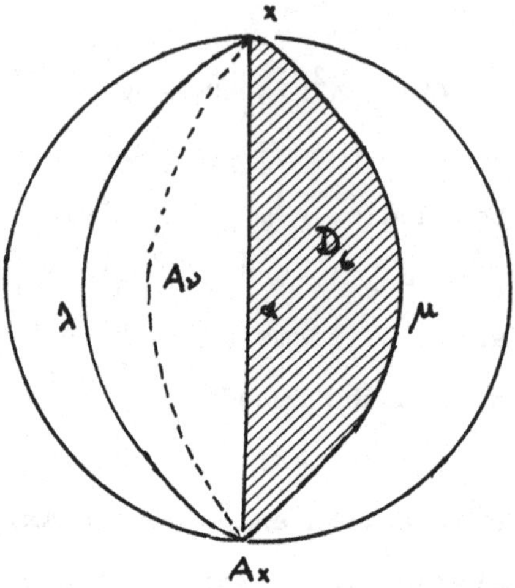

Fig. 3.

We can find an isotopy of P to P_o taking

D_1 to $D_6 \cup D_7$ (with boundary $\lambda \cup \mu$)

D_4 to $A^2D_6 \cup A^2D_8$ (with boundary $\nu \cup A^2\mu$), and

D_5 to $AD_7 \cup AD_8$ (with boundary $A\lambda \cup A^2\nu$),

keeping P fixed outside the union of the domain discs. Then for P_o we have

$$P_o \cap AP_o = D_6 \cup AD_7 \cup A^2D_8 ,$$

and by a final deformation in the neighbourhood of this intersection we obtain a purjective plane P_1 which is such that

$$P_1 \cap AP_1 = AP_1 \cap A^2P_1 = P_1 \cap A^2P_1 = \alpha \cup A\alpha \cup A^2\alpha = C_o .$$

If N_o is an A-invariant regular neighbourhood of C_o, then ∂N_o is mapped to a torus in the orbit space, for which the closures of the components of the complement are solid tori. This is enough to show that the orbit space is Seifert fibered and hence is homeomorphic to a lens space.

Recall that M^3 is a prism manifold if it is the orbit space of a free linear action by $C_u \times D^*_{4n}$ on S^3. Using the $(n_2,1)$ - Seifert decomposition of such a manifold explained in Chapter I we see that M^3 contains an embedded Klein bottle K^2. Furthermore, if we assume that u is odd and n is even,

$$H_1(M,\mathbb{F}_2) \cong \mathbb{F}_2 \times \mathbb{F}_2 .$$

THEOREM 3.3 *Let C_3 act freely on the manifold M just described. The orbit manifold either belongs to $\mathcal{M}$ or is Seifert fibred over S^3 with three exceptional fibres of multiplicities 2,3,3.*

Proof. As in 3.2 we start with the embedded Klein bottle K and its translates AK and A^2K in relative general position. If A generates C_3 then the induced action A on H_1 is either trivial or permutes the three elements of order 2. We concentrate on the second case, since it is this which we use in proving the reduction theorem (4.3) below. (The argument for the first is similar, see [R2] pages 168-70.) Here

$K \cap AK$ and $K \cap A^2K$ both contain one curve which is one-sided in K, and the two curves belong to distinct homotopy classes.

By applying cutting and glueing arguments similar to those in 3.2 we can suppose that the 1-sided component C_1 of $K \cap AK$ is such that $C_1 \cap A^2K = \emptyset$.

Next remove all the separating loops of $K \cap AK$. The contractible ones may be removed by the same argument as in 3.2. So suppose that $C \subset K \cap AK$ separates and is non-contractible in K. If N is the annulus lying between C and A^2C, then K-N consists of two Mobius strips E_1 and $E_2 (\partial E_1 = C, \partial E_2 = A^2C)$. By the assumption on H_1, either
$AK \cap \mathring{E}_1 = C_1$ & $A^2K \cap \mathring{E}_1 = \emptyset$, or
$AK \cap \mathring{E}_1 = \emptyset$ & $A^2K \cap \mathring{E}_1 = A^2C_1$.

Replacing K by $K_0 = (K - \mathring{E}_2) \cup A^2E_1$, and deforming this to be in relative general position with its translates, we obtain a Klein bottle with fewer separating loops in the intersection with its A-translate. Inductively we may suppose that all such have been removed.

Consider the one remaining curve C_1 in $K \cap AK$. Let N be a small tubular neighbourhood of C_1 such that $N \cap AN = \emptyset$. Write

$$R = M - \mathring{N} - (AN)^{\circ} - (A^2N)^{\circ} \text{ , so that}$$

$$R - K - AK - A^2K = \mathring{R}_1 \cup \mathring{R}_2 \text{ (disjoint union)}$$

when R_1 and R_2 are A-invariant solid tori. The orbit space is a union of 3 solid tori, which intersect in pairs along boundary annuli. As in the last part of the proof ofTheorem 3.1 this is enough to establish the existence of a Seifert fibration (with base space S^2 and at most 3 exceptional fibres). However knowledge of the fundamental group (Chapter II) plus the known list of fibrations (Chapter I) allows us to conclude that the exceptional fibres must have multiplicities 2,3 and 3.

In order to classify free actions on S^3 by the binary octahedral group $O_1{}^*$ and more generally to extend the proof of the reduction theorem (4.2) from Type III to Type IV it is necessary also to extend the argument of 3.3 from the cyclic group of order 3 to the dihedral group of order $2.3^s (s \geq 1)$, subject to the condition that some element of order 2 acts in a standard manner. The point here is that if $D(2.3^s)$ is acting on $M \in \mathcal{M}$ we can inductively arrange for only 3 of the 2.3^s translates of an embedded Klein bottle K to be distinct. Thus

THEOREM 3.4 <u>Let</u> $D(2.3^s)$ <u>act freely on</u> $M \in \mathcal{M}$, <u>and suppose that some element of order</u> 2 <u>has an orbit space in</u> $\mathcal{M}$. <u>Then either the orbit space for</u> $D(2.3^s)$ <u>belongs to</u> $\mathcal{M}$, <u>or is Seifert fibred over</u> S^2 <u>with</u> 3 <u>exceptional fibres of multiplicities</u> 2,3,4.

Proof. Let the dihedral group have the presentation

$$\{A, B : A^{3^s} = B^2 = 1, B^{-1}AB = A^{-1}\},$$

and suppose thatthe embedded Klein bottle K in M is invariant with respect to the subgroup $\langle A^3, B\rangle$. As before we may assume that K, AK and A^2K are in relative general position, and proceed to simplify the intersection curves. The main steps of the argument are:

1. $C = AC$ or $C \cap A^2K = \emptyset$ for each component C of $K \cap AK$. This is the hardest part of the argument, since we must allow for the $D(2.3^{s-1})$-action on K.

2. Use spherical modifications to replace K by a new Klein bottle K_o, such that no component of $K_o \cap AK_o$ is two-sided in K_o. This step is similar to the corresponding one in 3.3.

3. There are either one or two one-sided A-invariant curves remaining in $K \cap AK$. These can be used to construct a Seifert fibering of M, which is invariant under the group action. Two invariant curves give an orbit manifold in M, one the second possibility with three exceptional fibres. Again this follows from the general theory in Chapters I and II.

COROLLARY 3.5 Let $D(2^t.3^s)$, $t \geq 2$, $s \geq 1$ act freely on a manifold $M \in M$, and suppose that some non-normal subgroup C_2 has an orbit space belonging to M. Then the orbit space for $D(2^t.3^s)$ also belongs to M.

Proof. Consider the normal series

$$G_{t-1} \triangleleft G_{t-2} \cdots \triangleleft G_o = D(2^t.3^s),$$

in which $G_i \cong D(2^{t-1}.3^s)$ for $0 \le i \le t-1$ and $G_i/G_{i+1} = C_2$. Since the copy of C_2 in the assumptions of the corollary is not normal, it may be assumed to lie in G_{t-1}. Theorem 3.4 implies that M/G_{t-1} either belongs to M or Seifert fibres over S^2 with 3 exceptional fibres. However the structure of G_{t-2} eliminates the second possibility (which requires a group of binary octahedral type). Reworking the proof of Theorem 3.3 for free C_2 rather than free C_3-actions, which involves one rather than two translates of a Klein bottle, one can show that if M/G_i belongs to M, then so does M/G_{i-1}.

Notes and References.

The earliest and most original contribution to the theory of free topological actions by C_m on S^3 was made by G.R. Livesay in [Li]. His method was extended in an ad hoc way to cyclic groups of order equal to a low power of 2 by several authors, before R. Myers proved the generalisation presented here as Theorem 3.1. The essential homological argument goes over virtually without change from the pair (S^3,h) to the pair $(L^3(m,q),h)$, but gives a rather more complicated pair of invariant simple closed curves. Rubinstein's contribution (3.2-3.4) is quite independent, although there is overlap with Myer's approach if one replaces C_3 by C_2 see [R1]. However the property of possessing an embedded projective plane or Klein bottle is very restrictive, and no one has yet managed to extend the cutting and glueing arguments to one-sided surfaces of higher genus. For the theory of embedded surfaces, see the very interesting paper by G. Bredon & J. Wood, [B-W].

Using the results in this chapter one can come close to classifying free actions by finite groups of order $2^t.3^s$ on S^3. Surprisingly the groups not covered are the cyclic groups of order divisible by 3 (with the exception of C_6 and C_{12}). Recently J. Rubinstein has developed an entirely new method for dealing with these groups, which he calls the PL-minimax method. He uses this to prove that every free action by C_3 on a lens space is equivalent to a free linear action.

1. Consider one parameter families of embedded 2-spheres (for S^3) or embedded Heegard tori (for a lens space) which sweep out the total space of the group action. A typical family is a map

$$(S^2 \times [0,1],\ S^2 \times \{0,1\}) \to (S^3, \{x_1, x_2\}) \text{ or}$$

$$(T^2 \times [0,1],\ T^2 \times \{0,1\}) \to (L(p,q),\ \{c_1, c_2\})$$

2. Take such families which are in general position with respect to $C_3 = \{1, A, A^2\}$, that is, Σ_t , $A\Sigma_t$, $A^2\Sigma_t$ are in general position for all except a finite number of critical values of t. Here one has a single point at which the intersection is non-generic. For t non-critical count the number of triple points $\Sigma_t \cap A\Sigma_t \cap A^2\Sigma_t$, and let $m(\Sigma_t)$ be the maximum number.

3. The minimum value of the maximum $m(\Sigma_t)$ in (2) taken

over all one-parameter families is 6. (Note that the number of triple points is always divisible by 6 by orientability combined with the free C_3 action on the triple point set.) The key idea in the proof of this is to use the technique of Theorem 3.2 carefully to reduce the maximum for a family Σ_t if this exceeds 12. This is done by noting

(i) there must be a reasonable number of available 2-gons for each non-critical value of t, since one is dealing with three simple surfaces.

(ii) One must choose the 2-gons away from where critical points of the intersection of the three surfaces are about to occur, so as to continue with the family Σ_t after modification. Compare Pitts "Existence and regularity of minimal surfaces on Riemannian manifolds" Princeton Math. Notes, number 27(1981).

4. If the minimax equals 6 then the C_3 action is equivalent to a linear action. In the case of S^3 one can find an invariant unknotted S^1, and in the case of L(p,q) an invariant Seifert fibre, compare the last part of the proof of Theorem 3.1.

I am grateful to Rubinstein for sending me this summary (letter dated 20 September 1985).

CHAPTER IV: THE REDUCTION THEOREM

Using the results from the previous chapter it is easy to obtain the following classification.

THEOREM 4.1 <u>Let</u> G <u>be a finite group of order</u> $2^t.3^s$, <u>and if</u> G <u>is of Type</u> I <u>suppose that</u> s = 0 or 1, <u>then the orbit space</u> S^3/G <u>is elliptic</u>.

Proof. We consider the various types separately. If G is cyclic, that is $G \cong C_{2^t}$ or $C_{3.2^t}$, we apply Theorem 3.1 or Theorem 3.1 plus Theorem 3.2. If G is a non-abelian group of Type I, then because of the assumption on s, $G \cong D'(3.2^t)$ $(t \geq 2)$ contains a cyclic subgroup of index 2, and we may apply the result for cyclic groups plus 3.1.

Type II: $D^*(2^t)$, $D^*(2^t) \times C_{3^s}$, $D^*(2^t.3^s)$ with $t \geq 3$, $s \geq 1$. The first case again follows from 3.1. For the second consider the normal series

$$D^*(2^t) = G_0 \lhd G_1 \lhd \ldots G_s = D^*(2^t) \times C_{3^s} ,$$

where $[G_i : G_{i-1}] = 3$. Inductively we may suppose that G_{i-1} is the fundamental group of a manifold in $\mathcal{M}$, and apply 3.3. The induction starts, since by the previous step $S^3/G_0 \in \mathcal{M}$.

The proof in the third case is similar; $D^*(2^t.3^s)$ is an extension of a cyclic group of order 4 by the dihedral group $D(2^{t-2}.3^s)$. Since the group also contains a copy of the quaternion group of order 8, $D(2^{t-2}.3^s)$ contains a non-central subgroup C_2 such that the orbit space of the C_2-action on S^3/C_4 is in $\mathcal{M}$. We can now apply the corollary to 3.4.

Type III: T_v^* (apply 3.3).

Type IV: O_1^* (apply 3.4, noting that as in the third case of Type II the C_2- condition is satisfied).

Given the more recent work of Rubinstein on free C_3-actions (see the note at the end of the preceeding chapter) Theorem 4.1 actually holds for all groups of order divisible by 2 and 3 only. There are two additional groups to consider, $C_{2^t3^s}$ (s ≥ 2) to which the pℓ-minimax method applies directly and $D'(3^s2^t)$ (s ≥ 2), for which there is the normal series

$$D'(2^t3) \triangleleft D'(2^t3^2) \triangleleft \ldots D'(2^t3^s).$$

This allows us to argue as for Type II above. Perhaps as interesting is the next result, which reduces the general classification problem to one for cyclic groups of odd order, at least when G is solvable.

HYPOTHESIS 4.2 Let G be a finite group acting freely and

topologically on S^3. If $A \in G$ has odd order, assume that the action of G restricted to the subgroup <A> is conjugate to a free linear action in $Top^+(S^3)$.

This assumption is equivalent to supposing that the not necessarily regular covering space $S^3/_{<A>}$ is homeomorphic to a lens space.

THEOREM 4.3 _If_ G _is a group of Type_ I-IV _and satisfies_ 4.2 _then_ S^3/G _is elliptic_.

Proof. Again this proceeds case by case.
If G is cyclic, use 4.2 and Theorem 3.1. If G is a non-abelian group of Type I or II, G contains a cyclic subgroup of index 2 by the discussion in Chapter II. The result for cyclic groups together with 3.1 applies.
Now let G belong to Type III, $G = C_u^Y \times T_v^{*\,<X,A,B>}$, $(u,6) = 1$.

The index three subgroup generated by $\{Y,X^3,A,B\}$ is isomorphic to $C_{3^{v-1}\cdot u} \times D_8^*$, and by the previous step in the argument, the corresponding regular covering space is Seifert fibered. For example, unless $u = v = 1$ we may take orbit invariants $\{O;\ (n_2,1);\ (3^{v-1}u,2)\}$, and in the extreme case $\{2;\ (n_2,1)\}$. We have already observed that such a manifold contains an embedded Klein bottle, hence Theorem 3.2 applies to show that the covering transformation group is linear.

For Type IV present G as an extension

$$1 \to C_u \times D_8^* \to G \to D_6 \to 1.$$

Since a 2-Sylow subgroup G_2 is isomorphic to D_{16}^*, the table at the start of this section shows that the action restricted to G_2 is linear. Hence the additional condition on elements of order 2 in 3.4 and its corollary is satisfied, and the D_6 action on a manifold with orbit invariants $\{0;\ (n_2,1);\ (n,2)\}$ (say) is determined.

Remarks. If the proof of Theorem 3.1 can be generalised as suggested in the notes at the end of Chapter III, then the exceptional groups $Q(8;\ p,q)$ can be brought into the framework of Theorem 4.3. The group Q contains an index 2 subgroup of Type I (even an index 4 cyclic subgroup), which under Hypothesis 4.2 would correspond to a covering space admitting a Heegard decomposition of genus 2 (see J. Mennicke, Arch. Math. 8 (1957) 192-198), and the Appendix. An extension of Theorem 3.1 would then show that S^3/Q was Seifert fibered, contradicting the results in Chapter II. Note also that the same approach applies to Type IV groups in the theorem, since O_1^* contains T_1^* as a subgroup of index two.

In order to illustrate the restrictions which Seifert theory imposes in three dimensional topology let us show that there is no higher dimensional analogue of Theorem 4.3. Let D_{pq} be the non-abelian group of order pq, where p and q are distinct <u>odd</u> primes and $q|p-1$. Take the presentation

$$D_{pq} = \{A,B:\ A^p = B^q = 1,\ A^B = A^r,\ r^q \equiv 1 \pmod p\}.$$

THEOREM 4.4 <u>There exists a free action of</u> D_{pq} <u>on</u> S^{2q-1} <u>such that the</u> p- <u>and</u> q- <u>Sylow covering spaces, with fundamental groups generated by</u> A <u>and</u> B <u>respectively, are homeomorphic to lens spaces. However there exists no free linear action of</u> D_{pq} <u>on</u> S^{2q-1} .

Proof. A representation of D_{pq} over C can be decomposed into summands of the type

(i) β^k, $\beta(A) = 1$, $\beta(B) = e^{2\pi i/q}$, $0 \le k < q$, or

(ii) $\gamma_h = i_!\alpha^h$, $\alpha(A) = e^{2\pi i/p}$, $0 \le h < p$.

It is easy to see that the eigenvalues of γ restricted to A are $\{e^{2\pi mi/p}: m = 1,r,r^2...r^{q-1}\}$ and of γ restricted to B are $\{e^{2\pi ni/q}: n = 0,1,...,q-1\}$.

Given the B-eigenvalue equal to 1 there can be no free <u>linear</u> action on S^{2q-1} . However, since $q \ge 3$, $\dim(S^{2q}-1) \ge 5$ and we may use high dimensional techniques to construct free topological (even smooth) actions. Recall that one argues as follows [Th 6]; using a little more than the period 2q in cohomology, there exists a finite (2q-1)-dimensional complex P, which is homotopy equivalent to the desired manifold and which satisfies Poincaré duality. In order to perform surgery on some preimage of P embedded in a high-dimensional sphere we need an SO-bundle ξ over P and a map of degree one f: $S^{N+2q-1} \to T(\xi)$ (T denotes Thom space).

Since the covering spaces $P(p)$ and $P(q)$ corresponding to the subgroups $\langle A \rangle$ and $\langle B \rangle$ are certainly homotopy equivalent to lens spaces, which embed in the sphere, it is possible first to choose ξ to lift (stably) to the normal bundles ξ_p and ξ_q of these embeddings, and then, using the coprimeness of p and q, to find some map f of degree one into the Thom space. In this simple case we have used little more than the monomorphism

$$KO(P) \rightarrowtail KO(P(p))^{inv} \oplus KO(P(q)) ,$$

given for example by the Atiyah-Hirzebruch spectral sequence.

The middle dimensional surgery obstruction to replacing the complex by a manifold vanishes, because p and q are both odd. Furthermore because the Witt group of based Hermitian forms over the group ring ZD_{pq} injects into the Witt group over QD_{pq}, two such manifolds are homeomorphic if they have equal signatures (ρ) and Reidemeister torsions (Δ). Both invariants are detected by lifting to covering spaces corresponding to representative p- and q- Sylow subgroups.

Since $\langle B \rangle = G_q$ is a retract of $D_{pq} = G$ it is easy to satisfy the conclusion of the theorem at the prime q. For the prime p we note that the lift of ρ to G_p lies in the subset of $R(C_p^A)$ invariant under the action of C_q^B, and that the image contains lens space signatures. The lift of Δ is more subtle, because although it certainly lies in the invariant part of $K_1(QC_p^A)$, it is not a priori obvious that the image is large enough to contain the Reidemeister torsion of some lens space. This follows by an indirect geometric argument, due to T. Petrie [Pe]. Let $\gamma \oplus \beta$ act on the polynomial$\{\underline{z} | z_o^q + z_1^p + \ldots + z_q^p\}$. Then for suitable choices of ε and η

$$V_{\varepsilon,\eta}(q,p,\ldots,p) = S_\varepsilon^{2q+1} \cap \{\underline{z} | z_o^q + z_1^p + \ldots + z_q^p = \eta\}$$

is a Q-homology sphere, admitting a free D_{pq} action. (In essence freeness follows from the fact that $\gamma|\langle B\rangle$ has no more than one trivial eigenvalue, see above.) Taking a Q-cellular decomposition and using this to calculate the Reidemeister torsion, we see that it is possible to choose the lift of Δ at the prime p equal to $(A-1)(A^r-1)\ldots(A^{r^{q-1}}-1)$. Therefore the covering space for the prime p may be chosen to be homeomorphic to the lens space $L^{2q-1}(p;\, 1,r,\ldots r^{q-1})$.

Notes and references. Theorem 4.3 is no more than a collation of the results in Chapters two and three. As the summary of its proof shows Theorem 4.4 uses methods outside those developed in the notes, I have only included it in order to illustrate the very special character of dimension three - the classification theorem for D_{pq} actions is to be found in [Th 4], and 4.4 is an implicit corollary. For the background results and definitions, see the book of C.T.C. Wall [Wa], particularly Chapter XIV.

CHAPTER V: TANGENTIAL STRUCTURE

The aim of this section is to determine the images of the homotopy sets [BG, BSU(2)] and [BG, BSO(4)] in K(BG) and KO(BG) respectively, when G is a group with periodic cohomology. The result will then be combined with A. Hatcher's proof that the inclusion of SO(4) in $Diff^+(S^3)$ is a homotopy equivalence, and with the list of representations given in Chapter II, to show that if G is a group of Type I - IV acting freely on S^3, then S^3/G is homotopy equivalent to an elliptic manifold.

To begin with suppose that G is either cyclic of order p^t or binary dihedral, $D^*_{2^t}$. We shall stably describe [BG, BU(n)] and [BG, BSO(n)] starting from the isomorphisms

$$R(G)^\wedge = K(BG) = \lim_k K(BG^k)$$

$$RO(G)^\wedge = KO(BG) = \lim_k KO(BG^k),$$

proved for example in [At-S]. In both cases completion is with respect to the I-adic topology, where I is the kernel of the augmention map which sends a representation to its dimension. The isomorphism is obtained by completing the flat bundle homomorphism

$$\rho \longmapsto EG \times_G K^n, \; K = R \text{ or } C.$$

LEMMA 5.1 *If the group* G *has prime power order, the* I-*adic and* p-*adic topologies on the representation ring coincide*.

Proof. Consider the regular representation (real or complex) ρ with its underlying module KG of dimension p^n. Let x belong to the augmentation ideal, then $\rho x = 0$, since the character of ρ vanishes except at the identity, and the character of x vanishes at the identity. Therefore

$$(p^n - \rho)x = p^n x \text{ and } p^n I \subseteq I^2.$$

The other way round one uses the Adams operations to show that

$$\psi^{p^n}(x) = x^{p^n} + py, \text{ from which it follows that}$$

$$x^{p^n} = -py, \text{ since } \psi^{p^n}(x) \text{ vanishes.}$$

Thus $I^{p^n} \subseteq pI$, and the neighbourhood systems of the identity in RK(G) coincide.

THEOREM 5.2 *Let* $f\colon BG \to \begin{cases} BU(n) \\ BSO(n) \end{cases}, \; G \cong C_{p^t}$,

then there is a homomorphism $\phi\colon G \to \begin{cases} U(n) \\ SO(n) \end{cases}$, *such that the stable classes* [f] *and* [Bϕ] *coincide in* K(BG) or KO(BG)

Proof. Consider first the unitary case. Let k be any field of characteristic zero. If $a, b \in k$ interpret $(1 + at)^b$ as the formal binomial series

$$1 + bat + \frac{b(b-1)}{2} a^2t^2 + \ldots \in k[[t]].$$

LEMMA 5.3 Let a_i, $b_i \in k$ (a_i distinct and non-zero). If the formal power series

$$(1 + a_1t)^{b_1}(1 + a_2t)^{b_2}\ldots(1 + a_rt)^{b_r} = 1 + c_1 t + \ldots t \; c_n t^n \in k[t],$$

then each exponent b_i is a non-negative integer.

Proof. Having differentiated the given equation divide by it, obtaining

$$\frac{a_1b_1}{1 + a_1t} + \ldots + \frac{a_rb_r}{1 + a_rt} = \frac{a_1d_1}{1 + a_1t} + \ldots + \frac{a_rd_r}{1 + a_rt} + \frac{h'(t)}{h(t)} \; .$$

Here we have written the right hand side as $(1 + a_1t)^{d_1}\ldots(1 + a_rt)^{d_r} h(t)$, where $d_1 \ldots d_r \in Z$ and no $-1/a_i$ is a zero of $h(t)$.
Multiply in turn by the linear factor $(1 + a_it)$, and then set $t = 1/a_i$. The conclusion follows, since $a_ib_i = a_id_i$ and no $a_i = 0$. Now for the proof of 5.2.

Let ζ be a primitive p^tth root of unity and consider $G = C^A_{p^t}$. Let $\chi_A : R(G) \to C$ be the function which evaluates each simple character on the generator A; then after p-adic completion we have a map

$\chi_A^{\wedge} : R(G)_p^{\wedge} \to Z_p^{\wedge} \otimes C$ and the image takes values in the subdomain $Z_p^{\wedge}[\zeta]$ of the field $Q_p^{\wedge}[\zeta]$. In order to shorten the notation write $q = p^t$ and write $\xi, \xi^2, \ldots, \xi^q = 1$ for the simple representations of G: Suppose that $\chi_A(\xi) = \zeta$.

In the hypothesis of Theorem 5.2 write the class $[f] = b_1\xi + \ldots + b_q\xi^q$ in $Z_p^{\wedge} \otimes R(G)$, where each coefficient $b_i \in Z_p^{\wedge}$ and $\Sigma b_i \in Z$. If we write the total exterior power as

$$\lambda_t(x) = 1 + \lambda(x)t + \ldots + \lambda^r(x)t^r + \ldots, \text{ then}$$

$$\lambda_t[f] = (1 + \xi t)^{b_1}(1 + \xi^2 t)^{b_2} \ldots (1 + \xi^q t)^{b_q}.$$

But since the image of f is contained in BU(n), $\lambda_t f$ is a polynomial in t. After applying the completed character homomorphism $\lambda_A^{\wedge}$ we obtain the equation (in $Q_p^{\wedge}[\zeta]$ [[t]])

$$(1 + \zeta t)^{b_1}(1 + \zeta^2 t)^{b_2} \ldots (1 + \zeta^q t)^{b_q} = 1 + c_1 t + \ldots + c_n t^n.$$

Now apply Lemma 5.3, noting that $\xi, \xi^2 \ldots \xi^q = 1$ are distinct and non-zero.

If $G = C_{p^t}$ and we replace U(n) by SO(2n), the argument is similar except that we replace the field $Q_p(\zeta)$ by its subfield $Q_p^{\wedge}(\zeta + \overline{\zeta})$.

The argument just given almost extends to the binary dihedral groups $D^*_{2^s}$, thus we have

PROPOSITION 5.4 <u>Let</u> $f: BG \to \begin{cases} BU(n) \\ BSO(n) \end{cases}$, $G = D^*_{2^s}$, <u>then there are homomorphisms</u> $\phi_1, \phi_2: G \to \begin{cases} U(n) \\ SO(n) \end{cases}$, <u>such that the stable classes</u> $[f]$ <u>and</u> $[B\phi_1] - [B\phi_2]$ <u>coincide in</u> $K(BG)$ <u>or</u> $KO(BG)$.

Proof. Again first consider the unitary case, then the irreducible representations of $D^*_{2^s}$ are
$\{1, \alpha, \beta, \alpha \otimes \beta, \xi_m (1 \le m \le 2^{s-2}-1)\}$,

where the first three representations are one dimensional and factor through $C_2 \times C_2$, and as usual

$$\xi_m(A) = \begin{pmatrix} \eta^m & 0 \\ 0 & \eta^{-m} \end{pmatrix}, \quad \xi_m(B) = \begin{pmatrix} 0 & 1 \\ (-1)^m & 0 \end{pmatrix}.$$

Note that $\lambda_t(\xi_m) = \begin{cases} 1 + \xi_m t + \beta t^2, & m = \text{even} \\ 1 + \xi_m t + t^2, & m = \text{odd}. \end{cases}$

If using 5.1 we write $[f] = a_o 1 + a_1\alpha + a_2\beta + a_3(\alpha \otimes \beta) + \sum_m b_m \xi_m$,

with the coefficients a_i and b_m in $\hat{Z}_2$, then

$$\lambda_t[f] = (1+t)^{a_o}(1+\alpha t)^{a_1}(1+\beta t)^{a_2}(1+\alpha\otimes\beta t)^{a_3} \prod_{m_o=\text{even}} (1+\xi_{m_o} t + \beta t^2)^{b_{m_o}}.$$

$$\prod_{m_1=\text{odd}} (1+\xi_{m_1} t+t^2)^{b_{m_1}},$$

which is a polynomial since f maps into $BU(n)$. Applying the completed character functions $\hat{\chi}_A$, $\hat{\chi}_B$ and $\hat{\chi}_{AB}$ we obtain the expressions

$$\text{(a)}\quad (1+t)^{a_o+a_2}(1-t)^{a_1+a_3}\prod_m(1+2\cos m\theta+t^2)^{b_m} = \text{polynomial},$$

$$\text{(b)}\quad (1+t)^{a_o+a_1}(1-t)^{a_2+a_3}\prod_{m_o,m_1}(1-t^2)^{b_{m_o}}(1+t^2)^{b_{m_1}} = \text{polynomial},$$

and

$$\text{(ab)}\quad (1+t)^{a_o+a_3}(1-t)^{a_1+a_2}\prod_{m_o,m_1}(1-t^2)^{b_{m_o}}(1+t^2)^{b_{m_1}} = \text{polynomial},$$

and Lemma 5.3 now implies that

$$\text{(a)}\begin{cases} a_o+a_2 \\ a_1+a_3 \\ b_m \end{cases},\quad \text{(b)}\begin{cases} a_o+a_1+\Sigma b_{m_o} \\ a_2+a_3+\Sigma b_{m_o} \end{cases} \quad\text{and}\quad \text{(c)}\begin{cases} a_o+a_3+\Sigma b_{m_o} \\ a_1+a_2+\Sigma b_{m_o} \end{cases}$$

are all non-negative integers. Elementary linear algebra now implies that each coefficient a_i actually belongs to Z, which proves the theorem. Note that if one of the representations ξ_m occurs in [f], it has a positive coefficient (this could also be seen by restriction to a cyclic subgroup of index 2). However this is not the case for the one-dimensional representations - consider for example $-1+\alpha+\beta+\alpha\otimes\beta$ (corresponding to $a_o = -1$, $a_1 = a_2 = a_3 = 1$, $b_m = 0$), which is a class of virtual dimension 2 restricting to a homomorphism on each cyclic subgroup. The argument for $SO(n)$ is similar. We replace $R(D^*_{2^s})$ by $RO(D^*_{2^s}) = \{1, \alpha, \beta, \alpha \otimes \beta, \ldots\}$ and the field $Q_2^\wedge(\zeta)$ by $Q_2^\wedge(\zeta+\bar{\zeta})$. It is also necessary to observe that if $c \in Z$ and $c \in Z_2^\wedge$, then c is an integer because $\frac{1}{2} \notin Z_2^\wedge$.

Proposition 5.4 extends to groups of composite order as follows. Witt's induction theorem for real representations [Sw 2] Theorem 4.1, implies that a real class function is a virtual character if and only if this is the case after restriction to elementary and 2-hyperelementary subgroups. If G has periodic cohomology these subgroups H are either cyclic or (semi)direct products of a cyclic group C_r^A (r=odd) with $K = C_{2^t}^B$ or $D^*_{2^t}$. Write $\lambda: H \to K$ for the projection. Since the subgroup generated by A is normal, for each odd prime p, there is a unique Sylow subgroup $H_p \subseteq C_r^A$. If $f : BH \to BSO(n)$, let $[f_p]$ be the lift of the stable class of f to BH_p, by 5.3 $[f_p] = [B\phi_p]$ for some homomorphism ϕ_p, and the character of ϕ_p is invariant under the natural action of K on the characters of C_r^A. This holds because f is defined on BH rather than on BH_p. It follows as in [Se, 9.2] that the character $\chi(\phi_p)$ extends to a character χ_p on all of H; thus

$$\chi_p\,(A^s K) = \chi(A^{s'}),\ \text{where } s \equiv s' (\mathrm{mod}\ p^{t(p)})\ .$$

Define a global virtual character on H by

$$\chi = d + \sum_{p\ \mathrm{odd}} (\chi_p - d) + (\lambda^! \chi_2 - d)\ .$$

Here $d \leq n$ the virtual dimension of the class [f], and χ_2 is the possibly virtual character corresponding to f_2, the existence of which is guaranteed by 5.3 or 5.4.

If G is an arbitrary group with periodic cohomology, choose a representative family of Sylow subgroups $\{G_p\}$, and let $f_p : BG_p \to BSO(n)$ be the lift of $f : BG \to BSO(n)$. Since G_p is either cyclic or binary dihedral, 5.3 or 5.4 implies that f_p corresponds to a character χ_p, which since f_p is the lift of a global map f, satisfies

(i) $\chi_p(1) = d$, the virtual dimension over R of [f], and

(ii) $\chi_p(g_1) = \chi_p(g_2)$, whenever g_1 and g_2 are conjugate in G. Consider the class function

$$\chi(g) = d + \sum_p (\chi_p(g'_p) - d) ,$$

where g'_p is some element in G_p conjugate to the power of g generating the p-Sylow subgroup in the cyclic group <g>. Since χ restricts to the character χ defined above on each elementary or 2-hyperelementary subgroup H of G, Witt's induction theorem applies, and we have proved the harder part of

THEOREM 5.5 <u>Let</u> G <u>be a group with periodic cohomology and</u> $f : BG \to \begin{cases} BU(n) \\ BSO(n) \end{cases}$, <u>then there are homomorphisms</u> $\phi_1, \phi_2 : G \to \begin{cases} U(n) \\ SO(n) \end{cases}$, <u>such that the stable classes</u> [f] <u>and</u> $[B\phi_1] - [B\phi_2]$ <u>coincide in</u> K(BG) <u>or</u> KO(BG).

(The proof for maps into U(n) is slightly easier, since we can appeal to the Brauer rather than to the Witt induction theorem, and for an elementary subgroup we may use the formula $\chi = d + \sum_p (\lambda^!_p \ \chi_p - d)$.)

In the light of 5.2 and 5.5 it is possible to describe the image of the homotopy set [BG, BSU(2)] after stabilisation in K(BG) for each of the solvable groups G which admit fixed point free representations of (real) dimension four. Consider the diagram below:

$$\begin{array}{ccccc}
 & & \coprod_p \mathrm{Hom}(G_p, U) & & \\
 & & \big\downarrow \alpha_+ & & \\
\coprod_p [BG_p, BSU(2)] & \longrightarrow & \coprod_p K(BG_p) & \xrightarrow[c_2]{} & \coprod_p H^4(G_p, Z) \\
\uparrow \mathrm{Res} & & \uparrow \mathrm{Res} & & \uparrow \mathrm{Res} \\
[G, BSU(2)] & \longrightarrow & K(BG) & \xrightarrow[c_2]{} & H^4(G, Z) \\
 & & \big\uparrow \alpha_+ & & \\
 & & \mathrm{Hom}(G, U) \ . & &
\end{array}$$

The vertical arrows are all restrictions, natural with respect to both c_2 and stabilisation, α_+ is the flat bundle homomorphism, and we confine attention to positive representations. For each prime p the image of $[BG_p, BSU(2)]$ under stabilisation equals the image of α_+ restricted to homomorphisms into SU(2). When p is odd this is Theorem 5.2, when $p = 2$ and $G_2 = D^*_{2^s}$ we must look more closely at the proof of 5.4. In the notation used there

$$\text{(a)}\begin{cases} a_o + a_3 \\ a_1 + a_2 \\ \quad b_m \end{cases} \text{ are non-negative integers, and in addition}$$

$$a_o + a_1 + a_2 + a_3 + \sum_m b_m = 2 .$$

Hence at most one b_m can be positive, and in this case all other coefficients vanish. Assume therefore that

$$f \sim a_o 1 + a_1\alpha + a_2\beta + a_3(\alpha\otimes\beta), \text{ and}$$

$$a_o + a_1 + a_2 + a_3 = 2.$$

Conditions (a), (b) and (c) from 5.4 now imply that the sum of each pair $a_i + a_j$ $(i \neq j)$ is non-negative, and the analysis of 5.4 carried out for symplectic rather than real representations (SU(2) = Sp(1)), see [Th 5] shows that a_i is even $(0 \leq i \leq 3)$. Elementary linear algebra now shows that one value of a_i equals 2 and the others vanish. Therefore as in the case of cyclic groups $[f] = \alpha_+(\phi)$, for some homomorphism $\phi: D^*_{2^s} \to SU(2)$.

If G_p is cyclic, the second Chern class c_2 maps $[BG_p, BSU(2)]$ onto the subset of $H^4(G_p, Z)$ equal to $\{k^2 g_o \ , 0 \leq k < p^t\}$, where both here and elsewhere g_o is the generator corresponding to some reference fixed point free representation in SU(2). Similarly c_2 maps $[BD^*_{2^s}, BSU(2)]$ onto $\{k^2 g_o, 0 \leq k < 2^{s-1}\}$, the representations $2, 2\alpha, 2\beta, 2(\alpha \otimes \beta)$ make no contribution given the known structure of the cohomology ring, see for example [At].

For groups of composite order the easiest result to state is

PROPOSITION 5.6 *Let A_G be a maximal normal subgroup of odd order in the solvable group G, and let G admit some fixed point free representation in SO(4). Then, after stabilisation and restriction to A_G, each element in [BG, BSU(2)] is represented by a homomorphism $\phi: A_G \to SU(2)$. In particular the composition*

$$\text{Res}_{G \to A_G} \circ c_2 \text{ maps } [BG, BSU(2)] \text{ onto}$$

$$\{k^2 g_o , 0 \leq k < [A_G:1]\}$$

(In order not to overload the notation we write g_o both for the characteristic class of the chosen representation of G and for its restriction to A_G).

Proof. Since K(.) is a well-behaved cohomology theory the middle restriction map in the diagram following 5.5 is a monomorphism, the same is well-known to be true for H^4. If G is cyclic there are clearly enough representations to describe the image, and if G is a non-abelian group of Type I or II we can argue with a cyclic subgroup of index 2 containing A_G. If G is of type III or IV, the proof of Theorem 2.4 shows that G is defined by a short exact sequence

$$A_G \rightarrowtail G \twoheadrightarrow \begin{cases} T_1^* \\ O_1^* \end{cases},$$

where A is cyclic, and $G_p \subseteq A (p \geq 5)$. If $([A_G : 1], 6) = 1$, and this is always the case in Type IV, we can use the tensor product argument from the end of Theorem 2.6 to construct the necessary representations. If 3 divides the order of A_G, that is $G \cong C_u \times T_v^* \ (v \geq 2)$, G contains a subgroup of index 3 isomorphic to $A_G \times D_8^*$, and there are enough representations of T_v^* to map under c_2 to $\{k^2 g_o \ , \ 0 \leq k < 3^{v-1}\}$, see 2.6 again.

As a corollary of the argument and the description of $[BG_p, BSU(2)]$ for all primes (including 2), we have that if G is of Type I or II the conclusion of 5.6 holds with G replacing A_G. This is illustrated by the first of the following examples; the other two show that restriction to the subgroup A_G is necessary in Types III and IV.

EXAMPLES 1. $G = D^*_{4p}$ (p = odd prime, Type I). Let X generate a cyclic normal subgroup of order 2p, and let g_o be the characteristic class of the representation obtained by induction from the representation mapping X to $e^{\pi i/p}$. Replacing this 2p-th root by $e^{\pi k i/p}$, $0 \leq k < 2p$, we obtain a representation with characteristic class $k^2 g_o$ (p + 1 distinct values including 0 and $p^2 g_o$). Since the number of squares modulo p is $1 + (\frac{p-1}{2})$ and the number modulo 4 is 2, by 5.2 we have exhausted all possibilities.

2. $G = T_1^*$. If $f: BT_1^* \to BSU(2)$, the class of f in $K(BT_1^*)$ is represented by one of the following:

$$\{1, \omega \oplus \bar{\omega}, \tau, \psi^3(\tau)\}.$$

Furthermore, if $g_o = c_2(\tau)$, then $c_2[f] = k^2 g_o$ for $k = 0,1,3,4$.

For the representations we use the notation from Chapter II and ψ^3 is the non-stable Adams operation. The representation $\omega \oplus \bar{\omega}$ is the sum of the two non-trivial one dimensional representations of T^*_1 (and hence of T_1) which factor through $Z/_3^\times$. The only other possible value of k is 2, but $k^2 = 4$ (mod 8) is inadmissible by a symplectic argument, see [Th 5] again.

3. $G = O_1^*$. This is similar, and again we use the notation from Chapter II, with $c_2(O_+) = g_o$. The image of $[BO_1^*, BSU(2)]$ in $K(BO_1^*)$ is represented by $\{1, 2\rho, O_+, \psi^3(O_-), \lambda, O_-\}$ with characteristic classes $\{k^2 g_o, k = 0, 1, 3, 4, 5\}$. Here ρ is the one dimensional representation corresponding to the extra generator R, and $c_2(2\rho) = 0$ by the argument in [At] already referred to. The irreducible representation $\lambda: O_1^* \to SU(2)$ factors through D_6 (see [Ha] page 309), thus gives the class $16 g_o$. The classes $4g_o$ and $36g_o$ are again excluded, because on the subgroup

D^*_{16} ξ_2 is not symplectic.

Note that both in Examples 2 and 3, if $k^2 g_o$ is a generator of H^4 (i.e. $(k,6) = 1$) then $k^2 g_o$ corresponds to a homomorphism.

With these computations completed we can classify the free actions of G on S^3 up to homotopy type, still under the assumption that there is some free linear action to serve as "base point". Our argument depends on the proof of the Smale conjecture, given by A. Hatcher.

THEOREM 5.7 The inclusion homomorphism $i : SO(4) \longrightarrow Diff^+(S^3)$ is a homotopy equivalence.

Proof. See [Ht].

We remind ourselves that if G acts freely and linearly on S^3, then G is the direct product of cyclic group C_u with one of the model groups $D'_{2^t(2m+1)}$, D^*_{4m}, $T^*_v (v \geq 1)$ or O^*_1.

THEOREM 5.8 If G is one of the solvable groups just listed, and G acts freely and topologically on S^3, then the orbit space $S^3/_G$ is homotopy equivalent to an elliptic manifold.

Proof. There is no loss of generality in supposing that $S^3/_G$ is defined by a homomorphism $\rho: G \to Diff^+(S^3)$. (First triangulate the orbit space, and then observe that there are no obstructions to smoothing in dimension 3.) Next consider the diagram of classifying spaces.

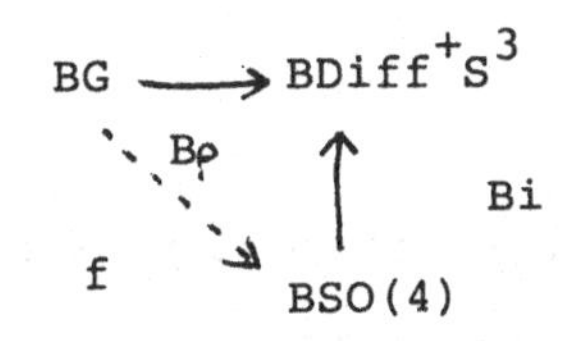

Hatcher's Theorem 5.7 implies that there is a factorisation $f\colon BG \to BSO(4)$ of $B\rho$, and by naturality the Euler class $e(f)$ equals the first k-invariant of $S^3/_G$. Recall that $e(Bf) = k_1(S^3/_G)$, since both are defined as primary obstructions, the latter to constructing a section of a $K(Z,4)$ fibration over BG, the former of the subfibration fibred by S^4. Furthermore the map $f\colon BG \to BSO(4)$ actually lifts to a map $\tilde{f}\colon BG \to BSpin(4)$, since $w_2(f) = 0$. This follows because $w_2(f)$ is detected by $H^2(G_2, F_2)$, and by Theorem 4.1 we already know that $S^3/_{G_2}$ is homeomorphic to a lens or prism manifold. In both cases we see that w_2 vanishes by looking at the trace of the matrices, which describe the defining free linear action. Hence because $Spin(4) \cong SU(2) \times SU(2)$ the homotopy class of the classifying map $B\rho\colon BG \to B\,Diff^+S^3$ is determined by a pair of maps $\tilde{f}_i\colon[BG, BSU(2)]$, $i = 1,2$.

We now argue type by type. If G is cyclic the result is classical, since there are enough fixed point free representations to cover all possible k-invariants. More generally the subgroup A_G is cyclic, and by the argument of 5.6 the stable class $[\tilde{f}|BA_G]$ is represented by $B\phi$, for some homomorphism ϕ. Because $\tilde{f}$ is defined on all of G, ϕ must be G/A_G invariant, and hence of the form $(\eta^k \oplus \eta^{-k})$ if G is non-abelian. It follows that, reduced modulo the order of A_G, the Euler class e(f) equals $k^2 g_o$.

If G is a group of Type I or II, G is a semidirect product of A_G and a 2-Sylow subgroup. If G is of Type III, G is the direct product of C_u and T_v^* , and A_G contains C_u as a subgroup of index 3^{v-1}. If G is of Type IV, $A_G = C_u$. With $H = A_G$(I,II & IV) or $H = C_u$(III) the results of Chapters III and IV imply that the action of G restricted to a subgroup isomorphic to $G/_H$ is conjugate in Diff^+S^3 to a free linear action. Hence, reduced modulo the order of $G/_H$, e(f) equals the Eulerclass of some free linear action. Inspection of the list of representations in Chapter II, given in more detail in Wolf's book, now shows that if we combine the values of e(f), modulo the orders of A_G and $G/_H$, there exists some representation $\rho_{lin} : G \to SO(4)$

such that $e(\rho) = k_1(S^3/_G) = e(\rho_{lin})$. This is enough to show that

$S^3/_G$ is homotopy equivalent to the elliptic manifold defined by ρ_{lin} .

Given the significance of cyclic groups of odd prime power order p^t it is worthwhile stating the slightly stronger version of 5.8 which holds for them.

COROLLARY 5.9 <u>If</u> $\rho: C_{p^t} \to \mathrm{Diff}^+S^3$ <u>defines a free action on</u> S^3, <u>then there exists a free linear action</u> $\rho_{lin}: C_{p^t} \to SO(4)$,

<u>such that the stable classes</u> $[B\rho_{lin}]$ <u>and</u> $[B(i^{-1}\rho)]$ <u>coincide in</u> $KO(BC_{p^t})$.

In some sense this says that the equivariant tangent bundle of S^3 cannot distinguish between linear and non-linear actions.

<u>Notes and references</u>

The results 5.2, 5.4 and 5.5 on maps between classifying spaces are special cases of results which hold much more generally. In the complex case and for am arbitrary compact Lie group G J. F. Adams has identified the image of the representation ring R(G) in K(BG), see [Ad]. If G is an arbitrary finite group (not necessarily with periodic cohomology), his results hold for RO(G) and RSp(G), where the latter is regarded as an RO(G)-module.

Theorem 5.8 can be regarded as a homotopy theoretic version of the reduction theorem 4.3. In essence it says that since $S^3/_G$ is homotopically linear for all cyclic subgroups, then the global action of G is also homotopically linear. The key ingredient here is Hatcher's Theorem, which like the results of Myers and Rubinstein in Chapter III is special to dimensions less than or equal to 3. The proof that I have given here is a simplified version of that in [Th 3].

It would be extremely interesting to "destabilise" the calculations for [BG, BSU(2)], even for cyclic groups of odd prime order. Unfortunately the results of H. Miller and A. Zabrodsky so far published are of no help, since they refer only to maps homotopic to a constant map. (Since BSU(2) (quaternionic projective space) has no multiplication, we cannot subtract two maps into BSU(2) which we wish to compare.) Until this homotopy set is better understood, we can do no more than use characteristic class arguments to show that some stable classes cannot factorise through BSU(2).

CHAPTER VI: $SL(2,F_5)$.

In this Chapter we consider the situation for the binary icosahedral group I^* (isomorphic to the special linear group of 2×2 matrices over the prime field F_5). Although it is plausible that any fixed point free homomorphism $I^* \to \mathrm{Diff}\ {}^+S^3$ is topologically conjugate to a homomorphism into $SO(4)$, the geometric picture in section three becomes more complicated. Thus, using the technique of J Rubinstein one is forced to consider five translates of an embedded K^2 rather than three (compare 3.3 - 3.5). Using an algebraic argument we shall show that under Hypothesis 4.2, the orbit space of an arbitrary free (smooth) action is at least simple homotopy equivalent to an elliptic manifold.

There are two irreducible representations of I^* in $SU(2)$, which induce free actions on S^3, denoted $\iota_\pm$ and interchanged by the unique outer automorphism of I^*. "Unique" here means unique up to composition with some conjugation!

PROPOSITION 6.1 *The Chern class* c_2 *maps the homotopy set* $[BI^*, BSU(2)]$ *onto the subset of elements in* $H^4(I^*,Z)$ *of the form* $\pm k^2g$, *where* $g = c_2(\iota_+)$ *and* $k^2 \not\equiv 4 \pmod 8$.

Proof. (see Ad , op. cit.) Since I* is a perfect group $(BI^*)^+$ is simply-connected and has the same cohomology. Since BSU(2) is also simply-connected, indeed as well-behaved as one could wish away from the prime 2, [BI*, BSU(2)] is determined by localisation at the primes 2, 3 and 5. The Sylow subgroups are respectively D_8^*, C_3 and C_5, so that by Theorem 5.2 and the discussion following Theorem 5.5 we have the following table for the possible values of k^2. Note that since we can always reverse the orientation it is enough to consider $k^2 < 60$.

k^2	0	1	4	9	16	25	36	49
mod 8	0	1	4	1	0	1	4	1
3	0	1	1	0	1	1	0	1
5	0	1	4	4	1	0	1	4
class	1	ι_+		$\psi^3(\iota_-)$		$\psi^5(\iota_+)$		ι_-

The columns corresponding to $k^2 = 4,36$ do not arise, since $[BD_8^*, BSU(2)] = [BD_8^*, BSp(1)]$, and symplectic considerations analogous to those used in chapter five show that the only possible Chern classes are $k^2 g_o$ with $k \equiv 0,1 \pmod 8$ see also [Th 5], Lemma 9.

COROLLARY <u>The first Pontrjagin class</u> p_1 <u>maps</u> $[BI^*, BSO(4)]$ <u>onto the subset</u> $\{\pm 2k g_o^2 \in H^4(I^*, Z) : k^2 \neq 4 (\mathrm{mod}\ 8)\}$.

Proof. Again apply the plus construction to BI^* and then localise. The stabilised image under restriction of $[BI^*_{(3\text{ or }5)}, BSO(4)_{(3\text{ or }5)}]$ contains only representations of the form $\eta^q \oplus \bar{\eta}^q$, because of invariance with respect to the normaliser of C_3 or C_5. Hence the corollary is true at odd primes. For the prime 2 note that $D_8^* \subset T_1^* \subset I^*$, and hence it is enough to consider the image of $[BT_1^*, BSO(4)]$ in $[BD_8^*, BSO(4)]$. The group T_1^* has only one irreducible (3-dimensional) representation which decomposes over D_8^* as the sum of non-trivial 1-dimensional representations; write this as $\alpha \oplus \beta \oplus (\alpha \otimes \beta)$. As complex representations suppose that $c.(\alpha) = 1 + a$, $c.(\beta) = 1 + b$ and $c.(\alpha \otimes \beta) = 1+a+b$. Taking account of the relations in $H^*(D_8^*, Z)$, see [At] page 62, we have

$$
\begin{aligned}
c.2(\alpha \oplus \beta \oplus (\alpha \otimes \beta)) &= [(1 + a)(1 + b)(1 + a + b)]^2 \\
&= [1 + a^2 + b^2 + ab + ab(a+b)]^2 \\
&= 1, \text{ since } a^2 = b^2 = 0.
\end{aligned}
$$

Arguing as in Proposition 5.4 we see that the stabilised image of $[BT_1^*, BSO(4)]$ in $KO(BD_8^*)$ is represented by homomorphisms rather than by virtual representations. (With the same notation as before it follows that $a_1 = a_2 = a_3$, $b \geq 0$ and $a_o + a_1 + a_2 + a_3 + 4b = 4$. If $b > 0$, $b = 1$ and all a_i's vanish; if $b = 0$, each a_i must equal one). The calculation of the Chern class of the complexified real representation above shows that the only non-zero contribution comes from $\xi \oplus \bar{\xi} \cong 2\xi$, and the corollary follows. Complement to the argument: if a class in $[BI^*, BSU(2)]$ has a Chern class which generates $H^4(I^*, Z)$, then stably it is represented by one of the homomorphisms $\iota_{\pm}$. This follows by inspection of the table above. Similarly the class p_1 maps $\iota_{\pm R}$ to generators of the subgroup $2H^4(I^*, Z)$.

PROPOSITION 6.2 <u>If</u> I^* <u>acts freely and smoothly on</u> S^3, <u>then the orbit space</u> S^3/I^* <u>is homotopy equivalent to an elliptic manifold</u>.

Proof. If the action is defined by a monomorphism $\phi: I^* \to \mathrm{Diff}^+S^3$, then again appealing to Hatcher's theorem that $SO(4)$ is a deformation retract of Diff^+S^3 we have a homotopy commutative diagram

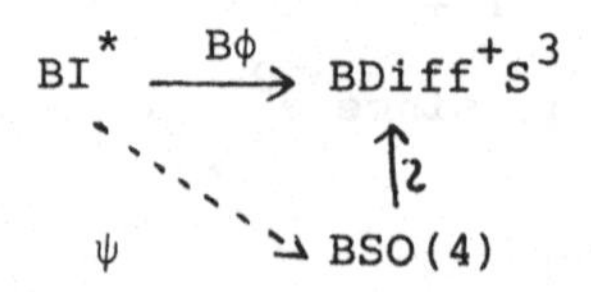

By naturality and obstruction theory $e(\psi) = k_1(S^3/I^*)$, and the Euler class is a generator of $H^3(I^*, Z)$. However ψ also factorises through $BSU(2)$; the obstructions are 2-primary and hence lift to a 2-Sylow covering space. This is homotopy equivalent to a prism manifold with fundamental group isomorphic to D_8^*, hence the listed map ψ_2 is homotopic to a map into BSU(2). At the odd primes 3 and 5 there is also factorisation through BSU(2) (rather than BU(2)) because of invariance with respect to the normalisers again. Hence we may interpret $e(\psi)$ as a stable second Chern class, and the complement to Proposition 6.1 shows that up to group automorphisms $k_1(S^3/_{I^*})$ equals the k-invariant of the unique elliptic manifold. We may now use obstruction theory once more to construct a homotopy equivalence.

The existence of a simple homotopy equivalence is more interesting. The orbit space Y^3 of a free smooth action by I^* on S^3 is a special complex in the sense of J.W. Milnor, see [Mi2] pages 404-411. Thus I^* acts trivially on the rational homology groups of S^3, and (at least up to homotopy) the equivariant chain complex in the universal cover has the form

$Z \rightarrowtail C_3(\tilde{Y}) \ \ldots \ \rightarrow C_o(\tilde{Y}) \twoheadrightarrow Z$, where C_3 and C_o are both free of rank one, and the subgroup of 3-cycles is generated by Σ, the sum of the group elements.

Working rationally let

$$QI^* = N \oplus (\Sigma), \text{ where}$$

N is the kernel of the homomorphism $QI^* \rightarrow Q$ which maps each group element to 1. N is itself an algebra with multiplicative identity equal to $1 - \Sigma/120$, and the subcomplex of $C_*(\tilde{Y},Q)$

corresponding to N is clearly acyclic and free. Hence there is a torsion element

$$\Delta(Y) = \tau\ (NC_*(\tilde{Y},Q))$$

belonging to the reduced Whitehead group of $N \cong QI^*/_{(\Sigma)}$.
Furthermore the general theory of Whitehead torsion shows that provided $SK_1(ZI^*) = \mathrm{Ker}(K_1(ZI^*)) \to K_1(QI^*))$ vanishes, the invariant $\Delta(Y)$ determines the simple homotopy type of the special complex Y. Thus

LEMMA 6.3 <u>A homotopy equivalence</u> $f: Y_1 \to Y_2$ <u>between special complexes is a simple homotopy equivalence if and only if</u> $f_*\Delta(Y_1)$ <u>equals</u> $\Delta(Y_2)$ <u>up to multiplication by</u> $\pm$ <u>(group element)</u>.

LEMMA 6.4 $SK_1(ZI^*) = 0$.

Proof. Regarded as a functor from the category of subgroups of a fixed finite group to abelian groups SK_1 satisfies a form of Frobenius reciprocity. It follows that $SK_1(ZI^*) = 0$ provided that $SK_1(ZH) = 0$ for each hyperelementary subgroup of I^*. Recall that H is hyperelementary if H is an extension of a cyclic group by a p-group of coprime order, and that for I^* each such subgroup us contained in a copy of D^*_{4p}, p = 2,3,5. Hence the lemma will follow from the more general assertion that $SK_1(D^*_{4p}) = 0$ if p is prime.

Suppose first that p is odd, and consider the three pull-back diagrams below:

$$D^*_{4p} = \{A,B\colon A^{2p} = 1,\ B^2 = A^p,\ A^B = A^{-1}\}$$

$$\begin{array}{ccc} ZD^*_{4p} & \xrightarrow{\langle B^2 - 1\rangle} & ZD_{2p} \\ {\scriptstyle \langle B^2+1\rangle}\downarrow & & \downarrow{\scriptstyle \langle 2\rangle} \\ W & \xrightarrow[\langle 2\rangle]{} & F_2D_{2p} \end{array}$$

$$\begin{array}{ccc} W & \xrightarrow{\langle 1+\bar{A}+\dots+\bar{A}^{p-1}\rangle} & H_- \\ {\scriptstyle \langle 1-\bar{A}\rangle}\downarrow & & \downarrow{\scriptstyle \langle 1-\zeta\rangle} \\ Z(j) & \xrightarrow[\langle p\rangle]{} & F_p(j) \end{array} \qquad \begin{array}{ccc} ZD_{2p} & \xrightarrow{\langle 1+A+\dots+A^{p-1}\rangle} & H_+ \\ {\scriptstyle \langle 1-A\rangle}\downarrow & & \downarrow{\scriptstyle \langle 1-\zeta\rangle} \\ ZC_2 & \xrightarrow[\langle p\rangle]{} & F_pC_2 \end{array}$$

In each case the ring in the upper left hand corner is split by dividing out by a pair of complementary ideals. $\bar{A}$ and A denote the images of A in W and ZD_{2p} respectively, j is the unit quaternion with $j^2 = -1$, and $H_\pm$ equals the twisted group ring $Z\widetilde{(\zeta)}\ (C_4^B)$ subject to the relations $B\zeta = \bar{\zeta}B$, $B^2 = \pm 1$, $\zeta = e^{2\pi i/p}$. Associated with each square there is a Mayer-Vietoris exact sequence of K-groups, and each ring in the lower right-hand corner has vanishing K_2. F_pC_2 and $F_p(j)$

are finite and semisimple, and by replacing Z with F_2 in the third square we can split F_2D_{2p} as F_2C_2 (K_2 vanishes by a calculation with 4 elements) and M_2 ($F_2(\zeta + \bar{\zeta})$) (K_2 vanishes for any finite field).

It follows that there is a monomorphism

$$K_1(ZD^*_{4p}) \to K_1H_+ + K_1(ZC_2) + K_1(H_-) + K_1(Z(j)).$$

H_+ and H_- are both hereditary orders, so by an argument first used by M. Keating, see [Ol 1], we may replace then by maximal orders. SK_1 is well-known to vanish for the remaining two rings and the lemma is proved.

The argument we have just outlined depends on being able to decompose the group ring in a nice way. For an alternative and much more general argument we refer to [Ol 2], where it is proved (Theorem 13.3) that $SK_1(ZG)$ is generated by the images under induction from elementary subgroups K of G. K is elementary rather than hyperelementary if K is the direct product of a cyclic group and a p-group of coprime order - and hence if G is a P-group of Type I, $SK_1(ZG) = 0$.

The vanishing of $SK_1(ZI^*)$ implies that the Whitehead group $Wh(I^*)$, which equals $K_1(ZI^*)$ modulo ($\pm$ image I^*) is torsion free. The rank is known to equal the number of irreducible real minus the number of irreducible rational representations, and the argument of 6.4 shows that the restriction map to $Wh(C_5)$ is non-trivial. The proof of the next theorem will show that the Reidemeister torsion $\Delta(S^3/I^*)$ is determined by that of a representative 5-Sylow covering space, and at least in the category of finite Poincare complexes there exist examples which are not simple homotopy equivalent to the lens space $L^3(5,-1)$. To see this we quote the result [Mi 2, page 375] that

$Wh(C_5^T)$ is infinite cyclic with generator $u = T + T^{-1}-1$.

Dividing C_5^T by the ideal generated by the sum of the group elements $\sum_{i=0}^{4} T^i$ gives the ring of algebraic integers $Z[\zeta]$, $\zeta = e^{2\pi i/5}$, for which K_1 is isomorphic to the group of units. Since 5 is a small prime, the torsion free summand of the group of units is generated by the cyclotomic unit

$$\frac{\zeta^2-1}{\zeta-1} = \zeta + 1 = u_1,$$

Furthermore if we consider the equivariant chain complex of S^3 corresponding to $L^3(5,-1)$,

$$Z \underset{\Sigma}{\rightarrowtail} ZC_5 \underset{T^{-1}-1}{\rightarrow} ZC_5 \underset{\Sigma}{\rightarrow} ZC_5 \underset{T-1}{\rightarrow} ZC_5 \underset{\varepsilon}{\twoheadrightarrow} Z ,$$

it is clear that Δ equals $(T^{-1}-1)(T-1) = -(T + T^{-1}-2)$, mapping to $u_2 = -(\zeta + \zeta^{-1}-2)$ in the ring $Z[\zeta]$. It follows that by using powers of u to twist the attaching map of the 3-dimensional cell in $L^3(5,-1)$ we can obtain finite Poincaré complexes of non-linear type. By using a unit in $ZI^*/(\Sigma)$ which restricts to at worst a power of u the same is true for the non-abelian group I^*. Hence the interest of

THEOREM 6.5 Let I^* act freely on S^3 and suppose that the action restricted to a representative 5-Sylow subgroup C_5 is conjugate to a free linear action. Then the orbit manifold S^3/I^* is simple homotopy equivalent to an elliptic manifold.

Proof. Theorems 3.1 and 3.2 imply that the action lifted to (a subgroup of) D_8^* or D_{12}^* is conjugate to a linear action. If we apply the argument of Theorem 3.1 to the inclusions

$$C_5 \hookrightarrow C_{10} \hookrightarrow D^*_{20}$$

we see that, given the assumption for C_5, the same is true for the third class of hyperelementary subgroup. Hence each hyperelementary covering manifold has a "linear" Reidemeister torsion - indeed since D_8^* and D_{12}^* are groups of exponent 4 and 6 respectively, their contributions to the global invariant Δ are trivial. This is therefore detected by Δ_{20}. A fixed point free representation of the binary dihedral group D_{20}^* takes the form

$$A \mapsto \begin{pmatrix} \zeta^q & 0 \\ 0 & \zeta^{-q} \end{pmatrix} \qquad B \mapsto \begin{pmatrix} 0 & 1 \\ -1 & 0 \end{pmatrix}, \quad \text{with}$$

$\zeta = e^{2\pi i/5}$ and $q = 1$ or 3. On restriction to the cyclic subgroup of order 5 generated by $A_1 = A^2$ we obtain the Reidemeister torsion Δ_5 equal to

$$(A_1^{-1} - 1)(A_1 - 1) \text{ or } (A_1^{-2} - 1)(A_1^2 - 1)$$

Since both representations extend to I^* giving ι_+ or ι_-, there is at worst an automorphism of I^* taking $\Delta(S^3/I^*)$ to the torsion invariant of an elliptic manifold. Lemma 6.4 now shows that the homotopy equivalence constructed in Proposition 6.3 can be taken to be simple.

At present we can only replace a simple homotopy equivalence by a homeomorphism by suspension to dimension 7, where the techniques of surgery theory used in Chapter 4 apply. Given a free action by I^* on S^3 defined by a homomorphism $\phi\colon I^* \to \mathrm{Diff}^+S^3$, we may construct a free (smooth) action on S^7, denoted by $\iota_+ * \phi\colon I^* \to \mathrm{Diff}^+S^7$, by taking the join of (S^3,ϕ) and the linear action (S^3,ι_+). Algebraically this amounts to splicing together two equivariant chain complexes across a copy of Z :

$$\cdots \to C_o(\iota_+) \dashrightarrow C_3(\phi) \to \cdots$$
$$\searrow \; Z \; \nearrow$$

Now in dimension $4k-1$, $k \geq 2$, a free action of I^* is classified up to homeomorphism by the following invariants: ρ, the multisignature in the real representation ring $RO(I^*)$, Δ, the Reidemeister torsion defined above for special manifolds, and ν_2, the mod 2 component of the normal invariant.

We have just seen how Δ is detected by restriction to hyperelementary subgroups; the same is true for ρ by the Witt induction theorem. The third invariant ν_2 is defined by a sequence of cohomology classes belonging to $H^{4\ell}(I^*, Z_{(2)})$ and $H^{4\ell+2}(I^*, Z/2)$, $1 \le \ell \le k - 1$, which in turn are detected by restriction to a 2-Sylow subgroup isomorphic to D_8^*. Therefore ν_2 is the normal invariant of an elliptic manifold. Since all three invariants are well-behaved with respect to the join construction we have

THEOREM 6.6 *If the free action on S^3 defined by the homomorphism ϕ is conjugate to a free linear action when restricted to a 5-Sylow subgroup, the action on S^7 defined by the join $\iota_+ * \phi$ is conjugate to a free linear action of I^*.*

Notes and References

The first part of this chapter isa reworking of material which has appeared in several places, exploiting the fact that I^* is perfect. The second part is standard algebraic K-theory, and is very clearly discussed in the survey article by J W Milnor. I would also recommend any reader interested in $SK_1(ZG)$ for an arbitrary finite group G to read the more recent survey article by R Oliver (available as of August 1985 as an Aarhus Universitet preprint). Under the assumption that all elements, and not just those of order 5, act linearly Theorem 6.4 extends weakly to direct products $C_u \times I^*$, where $(u,30) = 1$. Note that a maximal hyperelementary subgroup is now of the form $C_u \times D_{4p}^*$, with $p = 2,3$ or 5, belongs to Type I or II, and hence is associated with a "linear" Reidemeister torsion. The same holds for $C_u \times I^*$, perhaps after twisting by an automorphism of the form $\alpha_k \times \chi$, where $\alpha_k(u) = u^k$.

However $SK_1(Z(C_u \times I^*))$ need not vanish, see [Ol 2] Theorem 14.5, and so the homotopy equivalence is only weakly simple in the sense of Wall.

Theorem 6.6 depends on the calculation of the surgery obstruction groups $L_o^s(ZD_{4p}^*)$, contained in [Wa]. For an arbitrary odd prime p this contains both "local" and "finite adelic" terms, the latter vanishing if the class number of the pth. cyclotomic field is odd. Since this is certainly the case for p = 3 or 5 (the case p = 2 needs a separate argument), the multisignature ρ is enough to determine the orbits of the action of L_o^s on the set of manifolds with given Reidemeister torsion and normal invariant.

CHAPTER VII: FINITE POINCARE COMPLEXES AND HOMOLOGY SPHERES

So far we have considered free actions by finite groups on a single topological space - the 3-dimensional sphere with its standard differentiable structure. From the algebraic point of view this is very restrictive; the natural object to consider being a finitely dominated CW-complex homotopy equivalent to S^3. Any group G with cohomological period equal to 4 acts freely on such a space, and the number of distinct homotopy types equals the number of equivalence classes of generators of $H^4(G,Z)$ modulo the action of Aut G. As in Chapters V and VI we shall refer to such a generator as a k-invariant. If we now suppose that the complex on which G acts is finite rather than finitely dominated, the problems of existence and homotopy classification become much harder. In order to show what can be proved, and to point the difference with the theory for the very special space S^3, we shall consider the groups

$C_r, D_{2p}, D^*_{4p}, D^*_{2^t}$ and the three binary polyhedral groups

T^*_1, O^*_1 and I^* .

Here as always p is an odd prime number, and with the

exception of the dihedral group D_{2p} each group acts freely and linearly on S^3. Let $Y_o = S^3/G$ be some fixed reference orbit space with $g_o = k_1(Y) \in H^4(G,Z)$. Write $C_* = C_*(S^3)$ for the equivariant chain complex of the universal cover, and consider the exact sequence of finitely generated ZG-modules.

$$Z \rightarrowtail C_3 \to C_2 \to C_1 \to C_o \twoheadrightarrow Z.$$

The chain homotopy type of this complex can be varied as follows. Let r be coprime with the order G, and write (r,Σ) for the left ideal in ZG generated by r and the sum of the group elements Σ.

LEMMA 7.1. <u>The left ideals</u> (r,Σ) <u>are such that</u>

(i) $(r,\Sigma) + (r',\Sigma) \cong ZG + (rr',\Sigma)$

and

(ii) <u>there is an isomorphism</u> $\theta : Z + (r, \ ) \cong Z + ZG$ <u>such that the composition</u>

$$Z \rightarrowtail Z + (r,\Sigma) \cong Z + ZG \twoheadrightarrow Z$$

<u>is multiplication by</u> r.

Proof. (i) Choose s' and ℓ so that $r's' = 1 + \ell|G|$. The required map from right to left is given by $(1,0) \mapsto (u,s'u' - \ell v')$, $(0,u") \mapsto (r'u,\ell|G|u' - \ell r'v')$, $(0,v") \mapsto (v,0)$, where u and v generate the ideal (r,Σ) etc.

(ii) Choose s and ℓ so that $rs = 1 + \ell|G|$. The inverse isomorphism θ^{-1} is given by $\theta^{-1}(1,0)) = (s,\ell v)$, $\theta^{-1}(0,1)) = (1,u)$. Since θ^{-1} has degree equal to s (modulo $|G|$), θ has degree r (modulo $|G|$).

Property 1 implies that (r,Σ) is projective as a ZG-module of rank one, and that the collection of all such form a subgroup of the projective class group $\tilde{K}_o(ZG)$, which is usually denoted by T, the Swan subgroup. There is an alternative description in terms of the Mayer-Vietoris exact sequence associated with the diagram of rings

$$\begin{array}{ccc} ZG & \longrightarrow & ZG/(\Sigma) \\ \varepsilon\downarrow & & \downarrow \\ Z & \longrightarrow & Z/_{|G|} \end{array} .$$

The subgroup T equals the image of the boundary homomorphism with domain $K_1(Z/_{|G|}) = U(Z/_{|G|})$.

Using the isomorphism θ consider the exact sequence

$$Z + (r,\Sigma) \rightarrowtail C_3 + ZG \rightarrow C_2 \rightarrow C_1 \rightarrow C_o \twoheadrightarrow Z,$$

in which (r,Σ) is a Z-diect summand of $C_3 + ZG$ with Z-retraction $\overline{\phi}$. Because (r,Σ) is ZG-projective, there exists a homomorphism $\tau: (r,\Sigma) \to (r,\Sigma)$ such that $\Sigma\tau$ equals the identity, and $\phi = \Sigma(\tau\,\overline{\phi})$ is a ZG-retraction, see [C-E], Chapter XII, §1. Using any such retraction the sequence below remains exact

$$Z \rightarrowtail C_3 + ZG \underset{(d_3,\phi)}{\to} C_2 + (r,\Sigma) \underset{(d_2,0)}{\to} C_1 \to C_0 \twoheadrightarrow Z,$$

and the standard argument in homological algebra shows that there is a chain map between the initial and final complexes, which equals multiplication by r on the group of 3-cycles isomorphic to Z in both cases. Hence

LEMMA 7.2 *The generator* rg_0 *of* $H^4(G,Z)$ *corresponds to a finite* CW-*complex satisfying Poincare duality iff the projective module* (r,Σ) *is stably free*.

Note that by adding copies of ZG in dimensions 3 and 2 the stable finiteness of (r,Σ) is sufficient. Moreover there are no problems with geomatric realisation, since the original complex Y_0 is geometrically defined, and the 2-skeleton of the modified complex is obtained by adding a bouquet of copies of S^2.

Lemma 7.2 shows that for each of the group listed the number of distinct finite homotopy types depends on the structure of the Swan subgroup .

LEMMA 7.3 (i) $T(G) = 0$ *if* $G = C_r$ *or* D^*_{4p} (p = odd prime)

(ii) $T(D^*_{2^t})$ *is cyclic of order* 2, *in particular* $(r\ \Sigma) \sim 0$ *iff* $r \equiv \pm 1(8)$.

Proof. (i) If G is cyclic the quotient ring $ZG/_{(\Sigma)}$ is isomorphic to the ring of cyclotomic integers $Z[\zeta]$, and the map of units $U(Z[\zeta]) \to U(Z/_{|G|})$ is onto. By exactness T is trivial. Geometrically this corresponds to the fact that the generator qg_o of $H^4(G,Z)$ is realised by the lens space $L^3(r,q)$ for all values of q coprime with $|G|$.

If G is binary dihedral we use the same decomposition of the group ring as in Chapter VI. The class (r,Σ) is detected by its images over the rings ZC_2, H and $Z(j)$, where $j^2 = -1$. The argument for cyclic groups covers the first and last ($Z(j)$ is a quotient of ZC_4), and since Σ equals $(1 + A +\ldots+ A^{p-1})D$ the image of (r,Σ) in H is principal, hence free.

(ii) This is a variant of the first part. First note that if $r \equiv \pm 1 \pmod 8$, the existence of the prism manifolds shows that $(r,\Sigma) \sim 0$ in the projective class group. If $r \equiv \pm 3 \pmod 8$ we first restrict to the subgroup of $D^*_{2^t}$ generated by B and $A^{2^{t-4}}$ of order 8. A detailed calculation with units (which we omit) then shows that the summand $\{\pm 3\}$ in $U(Z/8)$ is not hit from behind in the appropriate Mayer-Vietoris exact sequence.

COROLLARY (i) <u>If</u> $G = D^*_{4p}$ (p = odd prime) <u>there exist finite</u> 3-<u>dimensional Poincaré complexes which are homotopically distinct from prism manifolds</u>.

(ii) *If Y^3 is a finite Poincaré complex with fundamental group isomorphic to* $D^*_{2^t}$, Y^3 *is homotopy equivalent to a prism manifold*.

Proof. The k-invariant of a prism manifold equals $\pm q^2 g_o$ see the discussion in Chapter II.

For each of the binary polyhedral group T^*_1, O^*_1 and I^* the Swan subgroup is detected by its restriction to hyperelementary subgroups, see [Sw2], which are either cyclic or binary dihedral of order 4p or 2^t. Hence Lemma 7.3 extends to them, by way of example, consider T^*_1. Since T^*_1 contains no subgroup of order 12, a hyperelementary subgroup is either cyclic or isomorphic to D^*_8. Therefore $(r,\Sigma) \sim 0$ if and only if $r \equiv + 1 \pmod 8$, and the stably free modules of this type are

$$(\pm 1,\Sigma) \text{ and } (\pm 17,\Sigma),$$

the first of which corresponds (up to orientation) with the unique free action of T^*_1 on S^3. If the standard chain complex $C_*(Y_o)$ is modified as in the proof of 7.2 by means of $(\pm 17,\Sigma)$ we again obtain a finite complex of exotic homotopy type. (Observe that an automorphism α of T^*_1 induces an automorphism of the unique 2-Sylow subgroup D^*_8, from which it follows that α^* is trivial in cohomology. Hence g_o and $17g_o$ do not belong to the same orbit in H^4.)

The argument in 7.2 can be extended to higher dimensions by splicing together copies of the chain complex $C_*(\tilde{Y}_o)$, and then modifying in dimensions $4k-i(i=1,2)$ so as to obtain a different homotopy type. For $k \geq 2$ there is no problem in replacing the modified complex by a homotopy equivalent manifold, since the surgery obstruction group $L_3(ZT_1*)$ vanishes. To see this again restrict to hyperelementary subgroups, and observe that $L_3(ZT_1*)$ is contained in the subgroup of $L_3(ZD_8*)$ left invariant under the action of the quotient group C_3. Once again direct calculation shows that this must be trivial.

In dimension 3 the vanishing of the surgery obstruction shows that the complex $\tilde{Y}^3$ can be replaced by a homology rather than by a homotopy sphere. That we should not expect this to be simply connected, and hence by Theorem 3.3 to be a counterexample to the Poincaré conjecture, is illustrated by the following construction.

The group T_1* has 3 irreducible complex representations of degree 2 which are non-trivial on the centre, and hence which do not factor through T_1. If we again consider T_1* as an extension to D_8* by a cyclic group of order 3 generated by X, each 2-dimensional representation restricts to the faithful representation $\xi = \xi_1$ on D_8*. Restriction to C_3^X gives the characters

$\chi(\tau = \tau_o) = \omega+\omega^2$, $\chi(\tau_1) = 1 + \omega$, $\chi(\tau_2) = 1 + \omega^2$, where ω is a primitive cube root of unity. The virtual representation

$$\rho = \tau_1 + \omega - 1$$

has augmentation 2, and restricts to the homomorphisms ξ and $\omega+\omega$ on the subgroups D_8^* and C_3^X respectively. If as usual

$$c_2(\tau) = g_o \in H^4(T_1^*, Z), \text{ then}$$

$$c_2(\rho) = 17g_o,$$

and ρ is a candidate for a homotopy class containing the stable classifying map of a homotopically exotic action.

In order to exploit this hint, recall the use of Brieskorn varieties in Chapter IV to construct free actions by the group D_{pq} on S^{2q-1}. Let $f(z_o, z_1)$ be some complex polynomial invariant under T_1^* acting via the representation τ_1, and let X act on z_2 by $z_2 \longmapsto \omega z_2$. Then for suitable values of ε and η

$$K(f) = \{\underline{z} \in C^3 : f(z_o, z_1) + z_2^3 = \varepsilon \ \& \ \underline{z} \in S^5(\underline{O}, \eta)\}$$

admits a free action of T_1^* induced by the representation $\tau_1 + \omega$. As in the case of D_{pq} the drop by one in the complex dimension corresponds to the elimination of a single eigenvalue equal to one. Since the direct sum of the equivariant tangent bundle of S^5 with a trivial real line bundle is determined by the homomorphism $\tau_1 + \omega$, it is not hard to see that the virtual representation introduced above $\tau_1 + \omega - 1$ determines $TK(f) \oplus (1)$. However it is also clear that we must expect $K(f)$ to have a large fundamental group, see for example the calculations in [O-W].

In high dimensions it is possible to use Brieskorn varieties in the same way as T. Petrie to obtain the homotopically exotic actions of T_1^* already mentioned. The non-vanishing middle dimension homology groups can be removed by surgery, since $L_3(ZT_1^*) = 0$. However it may be the case that a more elementary argument as in [Pe] is adequate.

Exercise. Apply Petrie's construction in dimension three with the dihedral group D_{2p} replacing D_{pq}. The complex surface

$$V(f_1) = \{\underline{z} \in C^3 : z_6^p + z_1^p + z_2^2 = \varepsilon\}$$

admits the D_{2p}-action given in terms of generators by

$$A\underline{z} = (\alpha z_o, \alpha^{-1} z_1, z_2) \text{ with } \alpha = e^{2\pi i/p}, \ B\underline{z} = (z_o, z_1, -z_2).$$

For suitable values of ε and η, the intersection $K(f_1)$ of $V(f_1)$ and $S^5(\underline{O}, \eta)$ admits an induced free action by D_{2p}.

Evaluate the fundamental group - given the famous theorem of J. Milnor on the non-existence of free dihedral group actions on spheres, this must be non-trivial. In higher dimensions 4k-1 the non-vanishing of the (2k-1)-dimensional homology groups can presumably be interpreted in terms of $L_3(ZD_{2p})$ by means of R. Lee's semicharacteristic, see[L].

Notes and References

The use of the modules (r,Σ) to vary the chain homotopy type of a periodic resolution is due to R. G. Swan, see [Sw3]. It is then easy to see that the Swan subgroup vanishes for cyclic groups; the calculations for the binary dihedral groups are by now familiar, and can be found in several places. The detection of $\tilde{K}_o(ZG)$ by hyperelementary subgroups is another example of the use of Frobenius reciprocity, see [Sw2]. I first came across the exotic T_1*-example when I was trying to use the first k-invariant to describe free actions on S^3 up to homotopy equivalence. Its existence shows that some geometric input is necessary in order to restrict the number of k-invariants which can occur. The discussion for the groups O_1* and I* is similar so far as finite Poincaré complexes go, but in high dimensions the construction of a homotopy equivalent manifold is more subtle, since the surgery obstruction group L_3 is no longer trivial. In dimension 3 the general method again gives free actions on homology spheres.

CHAPTER VIII: WORKPOINTS

The theory developed in the earlier chapters shows that the classification of free actions by finite groups on S^3 depends in an essential way on the special case of cyclic groups. In this chapter we propose to suggest various ways to approach the problem for cyclic groups of odd order, concentrating on prime powers p^t, $p \geq 5$. Our results are extremely partial; none should be regarded as more than a signpost to what may be true. We start with a result which is familiar to everyone who has worked in this area.

THEOREM 8.1 <u>Let the cyclic group</u> C_r <u>act freely and smoothly on</u> S^3. <u>Then if</u> S^3 <u>contains an embedded circle</u> S^1, <u>which is invariant and unknotted, the action of</u> C_r <u>is conjugate to a free linear action</u>.

Proof. Without a loss of generality we may suppose that C_r acts linearly on some tubular neighbourhood $S^1 \times D^2$ of the invariant circle, hence also on the boundary torus $S^1 \times S^1$. There is a second invariant circle S^1_2 embedded in the boundary, the position of which depends on the particular lens space $L^3(r,q)$ homotopy equivalent to S^3/C_r. Using a collar neighbourhood push S^1_2 into the complementary solid torus

$S^3 - (S^1 \times \mathring{D}^2)$ in such a way that it is still invariant. Again take a small tubular neighbourhood, and consider the action of C_r on the complement of the union of the two copies of $S^1 \times \mathring{D}^2$. In the orbit space we have an h-cobordism between two copies of $S^1 \times S^1$ (modulo a linear action). By an argument of J Stallings, see [St] steps 6-12, it is possible to use the Loop Theorem to trivialise the h-cobordism, so that (S^3, C_r) is obtained by identifying two linear actions on $S^1 \times D^2$ by means of an equivariant diffeomorphism of the common boundary. Up to isotopy, see for example [EE] this diffeomorphism can be taken to be the one defining the lens space $L^3(r,q)$, and the theorem is proved.

Theorem 7.1 shows that it is important to understand the class of knots in S^3 left invariant by some periodic diffeomorphism. Note that even in the case of linear actions, it is easy to produce invariant circles which are knotted; for example if C_2 acts via the antipodal map, take a standard S^1 and tie two complementary trefoil knots. The problem is to find some fixed point for the induced C_r-action on the space of unknotted embeddings of S^1 in S^3, modulo diffeomorphism of S^1. Note also that although the process of linearisation is particularly easy in the presence of a trivial knot, the argument of Theorem 3.1 shows that it is enough to produce an invariant knot of sufficiently simple type to allow the construction of a Seifert fibering of the space of orbits. We now illustrate this.

DEFINITION The knot K is said to have free period r if K is invariant with respect to a fixed point free diffeomorphism f of S^3 such that $f^r = 1$.

Such a diffeomorphism induces an automorphism of period r on the fundamental group πK of the complement. It is not hard to show that the torus knot of type (p,q) has period r for any value of r coprime with p and q, i.e. such a knot has infinitely many free periods. That this implication can be reversed is implicit in the work of W Thurston on hyperbolic 3-manifolds, see [T] and [F]. It also suggests that "highly divisible" group actions on S^3 are likely to have invariant circles which belong to a restricted class of knots.

DEFINITION The knot K is said to be a <u>satellite knot</u> if K is obtained from a non-trivial embedding of a knot K_1 in a small solid torus neighbourhood of K_2 .

Here non-trivial means that the embedding is neither isotopic to K_2 nor contained in a 3-disc embedded in the solid torus. If K is a satellite of K_2, K_2 is a companion knot of K. The Van Kampen theorem implies that

$$\pi K \cong \pi K_2 \underset{\pi_1(\partial V)}{*} \pi_1(V - K_1)$$

where V denotes the solid torus.

DEFINITION The free action by C_{p^t} on S^3 is said to be <u>infinitely</u> p-<u>divisible</u> if it extends to a free C_{p^n} action for all values of $n \geq t$. We shall denote such an infinitely divisible action by (S^3, C_{p^∞}).

THEOREM 8.2 <u>If the free C_{p^∞}-action on S^3 admits invariant circles, it extends to a free S^1-action, and is hence conjugate to a free linear action</u>.

Proof. An embedding of S^1 is C_{p^∞} invariant if it is invariant for all $n \geq t$. Although by general position it is clear that we can always find such an S^1 for any particular power p^n, without some restriction we may not be able to do so for all values of n simultaneously. Regarded as a knot K in S^3 this embedded S^1 has free periods of arbitrarily high order, and hence Aut(πK) is infinite. Since each automorphism can be realised by a self-homotopy equivalence of the manifold S^3-K, by the Mostow rigidity theorem it cannot have a hyperbolic structure. (The group of hyperbolic isometrics is finite.) Hence by the work of Thurston K is either a torus or a satellite knot. In the former case we are done, since we can use the method of 3.1 to give the orbit space S^3/C_{p^t} a Seifert fibre structure. Suppose therefore that K is a satellite knot, and let the free periodic map of S^3 induce the homeomorphism f on S^3-K. (If one wants to work with a compact manifold with boundary, replace S^3-K by S^3-(interior of a tubular neighbourhood of K).) S^3-K contains an embedded torus $\partial V'$ with

$$S^3 - K = M_1 \underset{\partial V'}{\cup} M_2 \text{ , where}$$

$fM_1 = V\text{-}K$ and $fM_2 = S^3 - \mathring{V}$ (V as in the definition of satellites). To see this we apply the loop theorem to do surgery on $U = f^{-1}(\partial V)$ as in [St]. It follows that if K has a free period r, then so does the companion knot K_2, and repeating the first part of the argument we see that K_2 is either a torus or a satellite knot. After finitely many steps of this kind we reach an invariant knot K_m, which is not a satellite, and such that $\text{Aut}(\pi K_m)$ is infinite. Therefore K_m must be a torus knot, and we are done.

Another line of attack is to attempt to construct a "good" Riemannian metric on the orbit manifold. This programme starts out well, but then runs into hard problems associated with the global integrability of partial differential equations. Let M^3 be a compact 3-manifold with Riemannian metric g_{ij}. By a first contraction of the curvature tensor associated with the Levi-Cività connection we obtain a symmetric tensor of degree 2, $\text{Ric}(g) = R_{ij}$, the Ricci curvature, and by a further contraction the scalar curvature S. M^3 is said to be an Einstein manifold if

$$R_{ij} = \lambda g_{ij}$$

for some constant λ, and it is not hard to see that such a manifold has constant sectional curvature, with universal covering spac S^3 or R^3. If the Ricci tensor is strictly positive it can also be shown that M^3 is the space of orbits of some free action by a finite group G on S^3, that is one of the manifolds we have been considering. In fact R Hamilton [Hm] has shown that such a manifold must be elliptic.

THEOREM 8.3 *Let* (M^3, g_o) *be a compact* 3-*manifold with positive Ricci curvature. Then there is a* 1-*parameter family of metrics* g_t, $0 \le t \le \infty$, *each with positive Ricci curvature, tending to a limit metric* g_∞ *of constant positive sectional curvature.*

Here are the main ingredients of the proof. By contraction of the Einstein condition we see that $\lambda = S/_3$, and hence it is natural to consider the flow g_t in terms of the heat equation

$$\frac{\partial g_t}{\partial t} = 2\,(1/3\; S_t g_t - \mathrm{Ric}(g_t))\,, \qquad (*)$$

subject to the initial condition $g_t|_{t-0} = g_o$. Replace this first partial differential equation by

$$\frac{\partial g_t}{\partial t} = 2\,(1/3\; \tilde{S}_t g_t - \mathrm{Ric}(g_t)), \qquad (**)$$

where (anticipating a final constant value S_∞) we replace S_t by its average value

$$\tilde{S}_t = \frac{1}{\mathrm{vol}\,(g_t)} \int_M S_t d\mu_t.$$

Consider a third equation

$$\frac{\partial g_t}{\partial t} = -2 \, \mathrm{Ric}(g_t) \qquad (***)$$

If g_t satisfies (***) define ψ_t so that $\psi_t g_t = \tilde{g}_t$ has volume one, and choose a new time variable $\tilde{t} = \int \psi_t dt$. Then $\tilde{g}_t$ satisfies (**) with t* replacing t. Having set up a suitable equation Hamilton proceeds as follows:

1. For any initial condition g_o (***) has a unique solution g_t defined on some interval $0 \le t < T$. The proof depends either on the Nash-Moser implicit function theorem, or on a further modification of the equation to make it strictly parabolic. This is necessary because

$$\mathrm{Ric} : S^2_+ (T^*M) \to S^2(T^*M)$$

is not strongly elliptic as an operator, because of invariance with respect to the group Diff(M).

2. For each time $t \in [0,T)$ the solution g_t has positive Ricci curvature. We start with the scalar curvature S_t, satisfying

$$\frac{\partial S_t}{\partial t} = \Delta S_t + 2|\mathrm{Ric}(g_t)|^2 \ .$$

This guarantees that $S_t > 0$ for all $t \in (0,T)$ provided that $S_o > 0$ - from the physical point of view the second term on the right may be considered to be a heat source, and an initially positive temperature cannot decrease with time. This argument can be adapted to the case when the solution function u_t takes values in some suitably ordered vector space, such as the space of self-adjoint operators on TM.

3. The solution metric g_t actually exists for all $t < \infty$. This part of the argument is technical and depends on a priori estimates.

4. The limit metric $g_\infty = \lim_{t\to\infty} g_t$ satisfies the Einstein condition. This is special to dimension 3, and is proved in two steps. For the first diagonalise the Ricci tensor as $\mathrm{diag}(\lambda_1, \lambda_2, \lambda_3)$, and then use the equation

$$3(\mathrm{trace}(\mathrm{Ric}^2) - \frac{(\mathrm{trace}(\mathrm{Ric}))^2}{3} = \sum_{1\le i<j\le 3} (\lambda_i - \lambda_j)^2 ,$$

which is identically equal to zero iff the eigenvalues converge to a common value. Taking bounds on the left hand side then shows that the common value is a positive constant. This second step concludes the argument.

The outline of the proof just given is based on a lecture given by R Palais at the 1982 Bonn Arbeitstagung, and the expository monograph [Kz], see Theorem 3.11.

The theorem above implies that in order to linearise a free

action C_r on S^3 it is enough to construct an invariant Riemannian metric of positive Ricci curvature on the standard 3-sphere. Using a homotopy equivalence

$$f : S^3/C_r \to L^3(r,q)$$

and defining $f^*R_{ij}(x) = R_{ij}f(x)$ it is possible to define a candidate for a Ricci tensor on S^3/C_r. Locally there are no problems in finding a solution to the essentially elliptic equation expressing f^*R_{ij} in terms of some metric g_{ij}, see [Kz] Theorem 3.4. Here local integrability depends on the fact that R_{ij} is invertible for at least one point, if we take it to be associated with the standard metric of positive curvature on the lens space. Since S^3/C_r is compact with finite fundamental group, the only known obstruction to global integrability vanishes see [Mi 3], pages 104-106. It may be the case that at least some of the obstructions are expressible in terms of the Reidemeister torsion Δ and the multisignature ρ, compare the discussion in the special case when r is prime below. (We avoid the Poincaré conjecture by assuming that the universal cover is always diffeomorphic to S^3.) Note that given recent work by J Lannes and others the final result in Chapter V on tangential type can be destabilised. Thus

THEOREM 8.4 <u>Let $\phi: C_p \to \mathrm{Diff}^+S^3$ define a free action by the cyclic group C_p of prime order on S^3. Then the factorisation of $B\phi$ through $BSO(4)$ is homotopy equivalent to the classifying map of a free linear action.</u>

Theorem 8.1 (topological) and Theorem 8.3 (geometric) outlined above give hints on a possible adaptation of high dimensional arguments to the special case of 3-manifolds with finite fundamental group. Recall the classification of fake lens spaces, i.e. orbit spaces of free C_p-actions on S^{2n-1}, $n \geq 3$, restricting attention to the case when p equals an odd prime number. Up to homeomorphism such actions are classified by Δ, the Reidemeister torsion, and ρ the (multi) signature. The former has been introduced in Chapter V, and takes values in the units of the ring $QC_p/(\Sigma)$, which are well-defined up to multiplication by plus or minus a group element. If T generates the cyclic group C_p

$$\Delta(L^{2n-1}(p;q_1,\ldots,q_n)) \sim (T^{q_1}-1)(T^{q_2}-1)\ldots(T^{q_n}-1)$$

In order to describe the latter invariant classify the C_p-action on S^{2n-1} by means of a map

$$S^{2n-1}/C_p \to K(C_p,1).$$

Since the bordism groups $\Omega^{top}_{2n-1}(C_p)$ are finite there exists a 2n-dimensional manifold W such that

$$\partial W = t(S^{2n-1}/C_p);$$

write $\tilde{W}$ for the covering manifold admitting a free C_p-action bounded by t copies of S^{2n-1}. Introduce the bilinear pairing

$$\phi : H^n(\tilde{W},R) \times H^n(\tilde{W},R) \to RC_p$$

$$(x,y) \longmapsto \sum_{i=0}^{p-1} (x \cup yT^{-i})T^i.$$

Then ϕ is a non-singular $(-1)^n$-Hermitian form over the group algebra RC_p, and if n is even, ϕ induces a symmetric form ϕ_i on each summand corresponding to a simple real representation R_i of C_p. (The complex group algebra CC_p is a direct sum of fields, so each R_i is either one or two dimensional.) Each form ϕ_i has a signature S_i and the collection of all such is called the (multi)signature.

Write $\sum_i s_i R_i = \text{sign}(\tilde{W},C_p) \in RO(C_p)$.

Now define $\rho(S^{2n-1}/C_p) = \frac{1}{t}\,\text{sign}\,(\tilde{W},C_p)$. Once more restricting n to be even it is not hard to show that ρ is well-defined as an element of

$$Q \otimes (RO(C_p) \text{ modulo the regular representation}).$$

There is an alternative definition of $\text{Sign}(\tilde{W},C_p)$ in terms of the formal difference between the positive and negative eigenspaces of $H^k(\tilde{W},R)$, at least intuitively it is clear that the two approaches give the same virtual real representation. Also by introducing a fixed point set into W, it is possible to remove the denominator t - thus the linear action on S^3 defining the lens space $L^3(p,q)$ extends smoothly to D^4 with a fixed point at the origin. In this way one can show that

$$\rho(L^3(p;q)) = \frac{(\tau+1)(\tau^q+1)}{(\tau-1)(\tau^q-1)},$$

where $\tau(T) = e^{2\pi i/p}$ for the chosen generator T of C_p, see [A-B].

An arbitrary free smooth action by C_p on S^3 extends topologically to D^4 with the origin as fixed point. If <u>this</u> action is smoothable the signature $\rho(S^3/C_p)$ equals the signature of the lens space defined by the representation of C_p on the normal space at the origin. However note that at least a priori, if the action is defined by the homomorphism $\phi: C_p \to \mathrm{Diff}^+S^3$, the invariant ρ is much more delicate than the homotopy class of the classifying map $[B\phi]$. Thus in dimension 9 rather than 3 there is an example of an orbit space with fundamental group of order 35, which is both simple homotopy and stably tangentially equivalent to the lens space, $L^9(35 ; 1,6,6,6,6)$ without being homeomorphic to it, see [Mi 2] page 410. Since by a generalisation of the classification theorem from C_p to $C_{p_1p_2}$ the invariants Δ and ρ still suffice, the signature must carry more information than $[B\phi]$. The present unsatisfactory state of our knowledge may be summarised in the following result, in which we suppose that the orbit manifold is polarised, i.e. there is a fixed identification of the fundamental group with the abstract cyclic group C_p^T.

THEOREM 8.5 Let the orbit space Y of a free linear action by C_p^T on S^3 be the lens space $L^3(p,q)$. Then

$$\Delta(Y) \sim (T^q - 1)(T - 1) \text{ and } \rho(Y) = \frac{(\tau+1)(\tau^q+1)}{(\tau-1)(\tau^q-1)},$$

Conversely if the invariants of the action take these values (S^3, C_p^T) embeds in a free action on S^7, which is conjugate to a linear action.

Proof. This is similar to the result in chapter six on the binary icosahedral group, and depends on the fact that Δ and ρ both behave well with respect to taking the join with some linear action on S^3.

In order to use this argument we must restrict ourselves to actions for which we can determine the invariants. Theorem 8.1 suggests that if an action is infinitely divisible then up to normalisation Δ and ρ take the values quoted in 8.1. In order to place restrictions of the values taken by the signature I suspect that we need a theory of stratified smoothings of stratified sets in dimension 3. The Reidemeister torsion looks more promising - there is some evidence from higher dimensions that comparatively mild divisibility assumptions imply that $\Delta(Y)$ equals $\Delta(L^3(p,q))$. As in Theorem 8.1 one can try embedding a given C_p-action in a free action by some cyclic group of large order, and then see what restrictions this places on Δ. We conclude with two questions:

QUESTION 1. *If* (S^3, C_p) *embeds in* (S^3, C_{rp}) *what are the possible values of the Reidemeister torsion* (a) *for* $r = 2^e$ *and* (b) *for* $r = p^e$.

QUESTION 2. *For each prime* p *is it possible to find some positive integer* e, *such that, if the free* C_p*-action on* S^3 *extends to a free* C_{2p^e} *- action, then* (S^3, C_p) *embeds in a free linear action on* S^7 ?

APPENDIX

(after J. Mennicke)

Genus two Heegard decompositions for elliptic manifolds.

In this appendix we shall show how to express each of the elliptic manifolds with non-abelian fundamental group as the union of two solid tori of genus 2 identified along the common boundary $(S^1 \times S^1) \# (S^1 \times S^1)$. The discussion in Chapter 2 has shown that such a manifold is determined up to homeomorphism by its fundamental group G, which belongs to one of five classes. As already suggested, this may be important in extending Myer s' results on free involutions or lens spaces to free involutions on prism spaces, and, for example, obtaining an alternative proof that free actions of O_1^* are conjugate to linear actions.

The homeomorphism of the two dimensional torus of genus 2 determining each manifold is itself determined by an automorphism of the fundamental group; let this be generated by a_1, a_2, b_1, b_2, where b_1 and b_2 are nulhomotopic in the solid tori. The group of automorphisms of the surface group

$\{a_1, a_2, b_1, b_2 : [a_1, b_1][a_2, b_2] = 1\}$

is generated by

$\sigma_1 = [a_1 b_1,\ b_1,\ a_2,\ b_2]$

$\sigma_2 = [a_1, a_1b_1, a_2, b_2]$ (1)

$\sigma_3 = [a_2, b_2, k_2^{-1}a_1k_2, k_2^{-1}b_1k_2]$, where $k_2 = a_2b_2a_2^{-1}b_2^{-1}$, and

$\sigma_5 = [a_1a_2b_2^{-1}, b_2a_2^{-1}b_1a_2b_2^{-1}, b_2a_2^{-1}b_1a_2b_2^{-1}b_1^{-1}a_2b_2a_2^{-1},$

$b_2^2\, a_2^{-1}\, b_1^{-1}\, a_2\, b_2^{-1}]$,

where the square bracket [, , ,] describes the images of a_1, b_1, a_2, b_2 under the automorphism α. A word in these four automorphisms determines a manifold, and Mennicke's calculations for the fundamental group are summarised in the table on the next page. The entries can be explained as follows: $\tau = \sigma_2^{x_1}\sigma_1^{y_1}\ldots\sigma_2^{x_n}\sigma_1^{y_n}$ &

$$\begin{pmatrix}1 & 0\\ x_1 & 1\end{pmatrix}\begin{pmatrix}1 & y_1\\ 0 & 1\end{pmatrix}\cdots\begin{pmatrix}1 & 0\\ x_n & 1\end{pmatrix}\begin{pmatrix}1 & y_n\\ 0 & 1\end{pmatrix} = \begin{pmatrix}p & q\\ f & g\end{pmatrix}. \quad (2)$$

Given the two positive coprime integers m, n we can solve the equations

$$m = -4f + 3g$$
$$n = -3f + 2g$$

for f and g in integers, thus

$$f = 2m - 3n$$
$$g = 3m - 4n\ .$$

Automorphism	Fundamental group
$\tau\ \sigma_3\ \sigma_2^{-1}\sigma_5^2\ \ \sigma_2^{-3}$	$f \equiv 0(2)\ \ C_{\lvert -4f+3g\rvert} \times D^*_{4\lvert -3f+2g\rvert}$ $f \equiv 1(2)\ \ C_{\frac{\lvert -4f+3g\rvert}{2^{t-2}}} \times D'_{2^t\lvert -3f+2g\rvert}$
$\tau\ \sigma_3\ \sigma_2^{-2}\sigma_5^2\ \sigma_2$	$g \not\equiv 0(3)\ \ C_{\frac{\lvert 2g+7\rvert}{3^{v-1}}} \times T^*_v$
$\tau\ \sigma_3\ \sigma_2^5\ \sigma_5^{-2}\sigma_2$	$g \not\equiv 0(3)\ \ C_{\lvert 4g+21\rvert} \times O_1^*$
$\tau\ \sigma_3\ \sigma_2^{-4}\sigma_5^2\ \sigma_2^{-1}$	$(f,10) = 1,$ $(f,g) = 1\ \ \ C_{\lvert -3f+40\rvert} \times I^*$

Since the matrices $\begin{pmatrix} 1 & 0 \\ 1 & 1 \end{pmatrix}$ and $\begin{pmatrix} 1 & 1 \\ 0 & 1 \end{pmatrix}$ generate the group of unimodular 2×2 integral matrices, given f and g we can determine the exponents x_i and y_i in τ from the matrix equation above. This shows for example in the first row, that if f is even, we obtain all groups of the form $C_m \times D^*_{4n}$. Similar numerical considerations show that we can choose f and g to exhaust all possible orders for groups of Types III, IV and V. For the last case, if $f \equiv 1,3,7,9 \pmod{10}$ we obtain $|-3f+40| = 30k \pm 7,\ 30k \pm 1,\ 30k \pm 11,\ 30k \pm 13$.

In order to obtain the fundamental group from the automorphism first note that since σ_1 and σ_2 operate on a_1 and b_1 the subword τ is described by the bracket

$$[\phi(a_1,b_1),\ \psi(a_1,b_1),\ a_2,b_2].$$

The complementary subword must then be written out in full, using the equations (1), giving $[w_1,v_1,w_2,v_2]$. Since the generators b_1 and b_2 are nulhomotopic in the solid torus, the relations in the fundamental group are

$$v_1(a_1,1,a_2,1) = v_2(a_1,1,a_2,1) = 1.$$

Furthermore the matrix product (2) shows that

$$\phi(a_1,1) = a_1{}^g\ ,\ \psi(a_1,1) = a_1{}^f\ .$$

An an illustration let G be of type I or II, associated with the automorphism $\tau\upsilon$, $\upsilon = \sigma_3\sigma_2{}^{-1}\sigma_5{}^2\sigma_2{}^{-3}$. Then

$$v_1 = (k_2^{-1}\psi\phi^{-1}k_2b_2a_2^{-1})^2(k_2^{-1}\phi\bar{\psi}^{-1}k_2a_2b_2^{-1})^2(b_2a_2^{-1}(a_2(b_2a_2^{-1})^2$$

$$(a_2b_2^{-1}k_2^{-1}\phi\psi^{-1}k_2)^2)^{-3}$$

$$v_2 = k_2^{-1}\psi(\psi\phi^{-1})^2k_2(k_2^{-1}\psi\phi^{-1}k_2b_2a_2^{-1})^{-2} .$$

Using the description of ϕ and ψ above, and setting $b_1=b_2=1$, we obtain

$$1 = (a_1^{f-g}a_2^{-1})^2(a_1^{g-f}a_2)^2a_2^{-1}(a_2^{-1}(a_2a_1^{g-f})^2)^{-3},$$

$$1 = a_1^{3f-2g}(a_1^{f-g}a_2^{-1})^{-2} .$$

It follows from the second relation that a_1^{3fg-2g} and a_2 commute. Replacing $(a_1^{f-g}a_2^{-1})^2$ by a_1^{3f-2g} in the first relation we obtain

$$1 = a_1^{3f-2g}\ a_1^{-3f+2g}\ a_2^{-1}(a_2^{-1}a_1^{-3f+2g})^{-3} .$$

The first two factors cancel, and expanding the bracket, we obtain

$$a_2^2 = a_1^{3(-3f+2g)} .$$

The second relation, multiplied on the left by a_2^{-2} gives

$$a_2^{-1}a_1^{-f+g}a_2 = a_1^{7f-5g} .$$

If we introduce the new generators $a_1' = a_1$, $a_2' = a_2a_1^{-f+g}$ we finally obtain the presentation

$$\{a_1' ,\ a_2' :\ a_2'^2 = a_1'^{-3f+2g} ,\ a_1^{12(-4f+3g)(-3f+2g)} = 1,$$

$$a_2'\ a_1'^{-f+g}a_2'^{-1} = a_1'^{7f-5g}\}$$

from which the entries in the right hand column of the table follow.

REFERENCES

[A-B] M. F. Atiyah - R. Bott, A Lefschetz fixed point formula for elliptic complexes II, Applications, Annals of Math., 88 (1968) 451-491.

[Ad] J. F. Adams, Maps between classifying spaces II, Inv. Math. 49 (1978) 1-65.

[At] M. F. Atiyah, Characters and cohomology of finite groups, Pub. math. IHES 9 (1961) 23-64.

[At-S] M. F. Atiyah - G.B. Segal, Equivariant K-theory and completion, J. Diff. Geom. 1 (1969) 1-18.

[B-W] G. Bredon - J. Wood, Non-orientable surfaces in orientable 3-manifolds, Inv. Math. 7 (1969) 83-110.

[C-E] H. Cartan - S. Eilenberg, Homological Algebra, Princeton University Press (1956).

[E-E] C. Earle - J. Eells, The diffeomorphism group of a compact Riemann surface, Bull. Amer. Math. Soc. 73 (1967) 557-559.

[F] E. Flapan, Ph.D. thesis (to appear).

[H] J. Hempel, 3-manifolds, Annals of Math. Studies 86 (1976) Princeton University Press.

[Ha] M. Hall, Theory of groups, (1959) Macmillan (London & New York).

[Hi] F. Hirzebruch - W. Neumann - S. Koh, Differentiable manifolds and quadratic forms, Lecture notes in pure and applied mathematics 4 (1971) Marcel Dekker (New York).

[Hm] R. Hamilton, Three-manifolds with positive Ricci curvature, J. Diff. Geom. 17 (1982) 255-306.

[Ho] H. Holmann, Seifertsche Faserräume, Math. Ann. 157 (1964) 138-166.

[Ht] A. Hatcher, A proof of the Smale conjecture, Diff $(S^3) \simeq O(4)$, Annals of Math. (2) 117 (1983) 553-607.

[J] K. Jänich, Differenzierbare G-Mannigfaltigkeiten, LN 59 (Springer Verlag, Heidelberg) 1968.

[Kz] J. Kazdan, Prescribing the curvature of a Riemann manifold, AMS regional conference series in mathematics 57 (1985).

[L] R. Lee, Semicharacteristic classes, Topology 12 (1973) 183-199.

[Li] R. Livesay, Fixed point free involutions on the 3-sphere, Annals of Math. 72 (1960) 603-611.

[Mi[1]] J. Milnor, Groups which act of S^n without fixed points, Amer. J. Math. 79 (1957) 632-630.

[Mi[2]] J. Milnor, Whitehead Torsion, Bull. Amer. Math. Soc. 72 (1966) 358-426.

[Mi[3]] J. Milnor, Morse Theory, Annals of Math. Studies 51 (1963) Princeton University Press.

[Mn] J. Mennicke, Über Heegarddiagramme vorn Geschlcht zwei mit endlicher Fundamentalgruppe, Arch. Math. 8 (1957) 192-198.

[My] R. Myers, Free involutions on lens spaces, Topōlogy 20 (1981) 313-318.

[O] P. Orlik, Seifert Manifolds, LN 291 (Springer Verlag, Heidelberg) 1972.

[O-W] P. Orlik - P. Wagreich, Seifert n-manifolds, Inv. Math. 28 (1975), 137-159.

[Pe] T. Petrie, Free metacyclic group actions on homotopy spheres, Annals of Math. 94 (1971) 108-124.

[R1] J. Rubinstein, On 3-manifolds that have finite fundamental group and contain Klein bottles, Trans. Amer. Math. Soc. 251 (1979) 129-137.

[R2] J. Rubinstein, Free actions of some finite groups on S^3, I, Math. Ann. 240 (1979) 165-175.

[Se] J. P. Serre, Représentations linéaires des groupes finis, 2e ed., Hermann (Paris, 1971).

[St] J. Stallings, On fibering certain 3-manifolds, Topology of 3-manifolds and related topics, 1962 (Prentice-Hall, NJ) 95-100.

[Sw[1]] R. G. Swan, The p-period of a finite group, Ill.J. Math. 4 (1960) 341-346.

[Sw[2]] R. G. Swan, Induced representations and projective modules, Annals of Math. 71 (1960) 552-578.

[Sw[3]] R. G. Swan, Periodic resolutions for finite groups, Annals of Math. 72 (1960) 367-291.

[T] W. Thurston, The geometry and topology of 3-manifolds, Princeton University notes.

[Th[1]] C. B. Thomas, On Poincaré 3-complexes with binary polyhedral fundamental group, Math. Ann. 226 (1977) 207-221.

[Th[2]] C. B. Thomas, Free actions by finite groups on S^3, Proceedings of symposia in pure mathematics, American Math. Soc. 32 (1978) 125-130.

[Th^3] C. B. Thomas, Homotopy classification of free actions by finite groups on S^3, Proc. London Math. Soc. (3) 40 (1980) 284-297.

[Th^4] C. B. Thomas, Classification of free actions by some metacyclic groups on S^{2n-1}, Ann. Sci. Éc. norm. sup., 4e serie 13 (1980) 405-418.

[Th^5] C. B. Thomas, Maps between real classifying spaces, Math. Ann. 253 (1980) 195-203.

[Th^6] C. B. Thomas, Obstructions for group actions on S^{2n-1}, LN 108 (Springer Verlag, Heidelberg) (1969) 78-85.

[Th-W] C. B. Thomas - C.T.C. Wall, On the structure of finite groups with periodic cohomology, Liverpool University preprint (1979).

[W] C.T.C. Wall, Surgery on compact manifolds, Academic Press (London 1970).

[Wo] J. Wolf, Spaces of constant curvature, 1st ed., McGraw Hill (New York 1967).

[Z] H. Zassenhaus, Über endliche Fastkörper, Abh. Math. Seminar Hamburg 11 (1936) 187-220.

Index